Multimedia Environmental Models

The Fugacity Approach

Donald Mackay

LEWIS PUBLISHERS

Library of Congress Cataloging-in-Publication Data

Mackay, Donald, 1925–
    Multimedia environmental models: the fugacity approach/Donald
Mackay.

        p.   cm.
    Includes bibliographical references and index.
    1. Organic compounds—Environmental aspects.   2. Pollution—
Mathematical models.   I. Title.
    TD196.073M33   1991   628.5′2—dc20   91-8600
    ISBN 0-87371-242-0

LEWIS PUBLISHERS, INC.
121 South Main Street, Chelsea, Michigan 48118

PRINTED IN THE UNITED STATES OF AMERICA

2   3   4   5   6   7   8   9   0

94   13719

For Ness, Neil, and Ian

TOXICOLOGY AND ENVIRONMENTAL HEALTH SERIES

Series Editor: Edward J. Calabrese

PREFACE FOR THE SERIES

Given the complex and ever-expanding body of information in toxicology and environmental health, the purpose of the Toxicology and Environmental Health Series is to present a genuine synthesis of information that not only will offer rational organization to rapidly evolving developments but also will provide significant insight into and evaluation of critical issues. In addition to its emphasis on assessing and assimilating the technical aspects of the field, the series will offer leadership in the area of environmental health policy, including international perspectives. Thus, the intention of this series is not only to provide a careful and articulate review of critical areas in toxicology and environmental health but to influence the directions of this field as well.

The Editorial Board will oversee and shape the series, while individual works will be peer-reviewed by appropriate experts in the field.

Edward J. Calabrese
University of Massachusetts
Amherst, Massachusetts

Donald Mackay, born and educated in Scotland, received his degree in Chemical Engineering from the University of Glasgow. After a period of time in the petrochemical industry he joined the University of Toronto, where he is now a professor in the Department of Chemical Engineering and Applied Chemistry, and in the Institute for Environmental Studies. Professor Mackay's primary research is the study of organic environmental contaminants, their sources, fate, effects, and control, and particularly in understanding and modeling their behavior with the aid of the fugacity concept. His work has focused especially on the Great Lakes Basin and on cold northern climates.

CONTENTS

PREFACE

This book is about the behavior of chemicals, primarily organic chemicals, in our multimedia environment or biosphere of air, water, soil, sediments, and the diversity of biota which reside in these media. It is a response to the growing concern that we have unwisely contaminated our environment with a large number of chemicals in the mistaken belief that its enormous capacity to dilute and degrade will reduce concentrations to negligible levels. It is now clear that the environment has only a finite capacity to dilute and degrade. Certain chemicals can persist and accumulate to levels which have adverse effects on wildlife and even humans. Some chemicals have the potential to migrate from medium to medium, reaching unexpected destinations in unexpectedly high concentrations. There is thus a need to understand these processes, not only qualitatively in the form of assertions that DDT evaporates and bioaccumulates, but quantitatively as statements that DDT in a particular region evaporates at a rate of 100 kilograms per year and bioaccumulates from water at a concentration of 1 ng/L to fish at levels of 1 μg/g.

Whereas there are many excellent books on the behavior of chemicals in the single-medium environments of air, or water or soil, this book takes a broader approach of viewing chemical behavior in the total biosphere of connected biotic and abiotic compartments. Environmental science has learned that chemical behavior in this complex assembly of media is not a random process like leaves blowing in the wind. The chemicals behave in accordance with the laws of nature which dictate chemical partitioning tendencies and rates of transport and transformation. Most fundamentally, the chemicals are subject to the laws of conservation of mass, or the "mass balance." These constraints provide the opportunity to establish the quantitative expressions of chemical fate which society now demands. This book is largely concerned with developing and applying these expressions in the form of mathematical statements or "models" of chemical fate.

Two features of modern environmental science are particularly noteworthy in facilitating the quantitative treatment of chemical fate in our multimedia environment.

First is the remarkable reduction in cost and increase in availability of computers, resulting in an explosion in computer literacy. Many environmental calculations are complex and repetitive and are thus particularly suitable for implementation on computers. Accordingly, for many of the calculations described in this book, computer programs are provided with which many chemicals can be readily assessed in a multitude of environments. A diskette is provided giving several programs.

Second, is the concept of fugacity, which was introduced by G. N. Lewis in 1901 as a criterion of equilibrium and has proved to be a very convenient and elegant method of calculating multimedia equilibrium partitioning. It has been widely used in chemical processing calculations and shows promise as being even more valuable for calculating the environmental fate of chemicals. It is introduced in this book as a convenient and elegant method of explaining and deducing the environmental fate of chemicals.

This book has been written as a result of the author teaching a graduate-level course at the University of Toronto on "Environmental Pathways." It is hoped that it will be suitable for other graduate courses, and for practitioners of the environmental science of chemical fate in government, industry, and the private consulting sector. The simpler concepts are entirely appropriate for undergraduate courses, especially as a means of promoting sensitivity to the concept that chemicals, which provide modern society with so many benefits, must also be more carefully managed from their cradle in the chemical synthesis plant, to their grave of ultimate destruction.

We can no longer regard the environment as a convenient, low cost, dumping ground for unwanted chemicals. If we are to discharge chemicals into the environment, it must be with a full appreciation of their ultimate fate, how they will be transported and transformed, and where and to what extent they may accumulate. We must ensure that mistakes of the past with PCBs, mercury, and DDT are not repeated. This is best guaranteed by building up a quantitative understanding of chemical fate in our total multimedia environment. It is hoped that this book is one step toward this goal and will be of interest and value to all those who value the environment and seek its more enlightened stewardship.

Multimedia Environmental Models

The Fugacity Approach

1   INTRODUCTION

Since the Second World War there has been growing concern about contamination of the environment by "man-made" chemicals. These chemicals may be present in industrial effluents, in consumer or commercial products, in mine tailings, or in petroleum products and gaseous emissions. Some chemicals such as pesticides may be specifically designed to alter the assembly of biota present in natural or agricultural ecosystems. They may be organic, inorganic, metallic, or radioactive in nature. Many are present naturally, but usually at much lower concentrations than have been established by human activity. Most of these chemicals cause toxic effects in organisms, including humans, if applied in sufficiently large doses or exposures. They may therefore be designated as "toxic substances."

There is a common public perception and concern that when these substances are present in air, water, or food, there is a risk of adverse effects to human health. Assessment of this risk is difficult because the exposure is usually (fortunately) well below levels at which lethal toxic effects, and even sublethal effects, can be measured with statistical significance against the "noise" of natural population variation and the simultaneous multiple toxic influences of other substances, some taken voluntarily and others involuntarily. There is a growing belief that it is prudent to ensure that the functioning of natural ecosystems is unimpaired by these chemicals, not only because ecosystems have inherent value, but because they can act as sensing sites or early indicators of possible impact on human well-being.

Accordingly, there has developed a branch of environmental science concerned with describing, first qualitatively, then quantitatively, the behavior of chemicals in the environment. This science is founded on earlier scientific studies of the condition of the natural environment – meteorology, oceanography, limnology, hydrology, and geomorphology and their physical, energetic, biological, and chemical subsciences. This newer branch of environmental

1

science has been variously termed "environmental chemistry," "environmental toxicology," or "chemodynamics."

It is now evident that if we are to exert effective and economic controls over the use of such chemicals we must have available methods of calculating their environmental behavior in terms of concentration, persistence, reactivity, and partitioning tendencies between air, water, soils, sediments, and biota. Such calculations are useful when assessing or implementing remedial measures to treat already-contaminated environments. They are useful for predicting the likely behavior of chemicals which may be newly introduced into commerce, or which may be subject to production increases, or introduction into new environments.

In response to this social need, this book develops, describes, and illustrates a framework and procedures for calculating the behavior of chemicals in the environment. It employs both the conventional procedures which are based on manipulations of concentrations, and procedures using the concepts of activity and fugacity to characterize equilibrium between environmental phases such as air, water, and soil. Most emphasis is on organic chemicals which are fortunately more susceptible to generalization than metals or inorganic chemicals when assessing environmental behavior.

The concept of fugacity, which was introduced by G. N. Lewis in 1901 as a more convenient thermodynamic equilibrium criterion than chemical potential, has been widely used in chemical process calculations. Its convenience in environmental chemical equilibrium or partitioning calculations has become apparent only in the last decade. It transpires that fugacity is also a convenient quantity for describing mathematically the rates at which chemicals diffuse, or are transported between phases; for example, volatilization of pesticides from soil to air. The transfer rate can be expressed as being driven by, or proportional to, the fugacity difference which exists between the source and destination phases. It is also relatively easy to transform chemical reaction, advective flow, and nondiffusive transport rate equations into fugacity expressions, and build up sets of fugacity equations describing the quite complex behavior of chemicals in multiphase, nonequilibrium environments. The equations adopt a relatively simple form which facilitates solution and interpretation of the dominant environmental phenomena.

We develop these mathematical procedures from a foundation of thermodynamics, transport phenomena, and reaction kinetics. Examples are presented of chemical fate assessments in both real and evaluative multimedia environments at various levels of complexity and in more localized situations such as at the surface of a lake.

These calculations of environmental fate are often tedious and repetitive, thus there is an incentive to use the computer as a calculating aid. Accordingly, computer programs are presented for many of the calculations described later in the text. It is important that the computer be viewed and used as merely a rather fast and smart adding machine, and not as a substitute for understanding. The reader is encouraged to write his or her own programs and modify those provided.

In preparing this material, I have benefited greatly from numerous texts, reports, and scientific papers. In particular the thermodynamic or phase equilibrium and physical chemistry aspects have relied heavily on the text *Molecular Thermodynamics of Fluid Phase Equilibria,* Second Edition by Prausnitz et al. (1986) and *The Properties of Gases and Liquids,* Fourth Edition by Reid et al. (1987). For interphase transport, the texts *Mass Transfer* by Sherwood, Pigford, and Wilke (1975), and *Transport Phenomena* by Bird, Stewart, and Lightfoot (1960) have been invaluable. In treating environmental aspects I have relied on *Chemodynamics* by Thibodeaux (1979), *Chemicals in the Environment* by Neely (1980), *Environmental Exposure from Chemicals* by Neely and Blau (1985), *Aquatic Chemistry* by Stumm and Morgan (1981), *Chemical Concepts in Pollutant Behavior* by Tinsley (1979), *Environmental Risk Analysis for Chemicals* by Conway (1982), *Pollutants in a Multimedia Environment* by Cohen (1986), and on the various issues of the *Encyclopedias of Environmental Chemistry* edited by Hutzinger.

It is a pleasure to acknowledge that much of the credit for the approaches devised in this book is due to the pioneering work of George Baughman, who saw most clearly the evolution of multimedia environmental modeling as a coherent and structured branch of environmental science amid the often frightening complexity of the environment and the formidable number of chemicals with which it is contaminated.

Finally, I am deeply indebted to my colleagues at the University of Toronto, especially Wan Ying Shiu and Sally Paterson, whose collaboration has been crucial in developing the fugacity approach, to Vivian Collier of Lewis Publishers for editorial assistance, and to my wife and Kartini Rivers for meticulous and cheerful preparation of the manuscript.

2   SOME BASIC CONCEPTS

2.1 INTRODUCTION

Much of the scientific fascination of the environment lies in its incredible complexity. It consists of a large number of phases such as air, soil, and water which vary in properties and composition from place to place (spatially) and with time (temporally). It is difficult to assemble a complete, detailed description of the condition (temperature, pressure, and composition) of even a small environmental system or microcosm consisting, for example, of a pond with sediment below and air above. It is thus necessary to make numerous simplifying assumptions or statements about the condition of the environment. For example, we may assume that a phase is homogeneous, or it may be in equilibrium with another phase, or it may be unchanging with time. The art of successful environmental modeling lies in the selection of the best, or "least-worst" set of assumptions which yields a model which is not too complex, yet is sufficiently detailed to be useful. The excessively simple model may be misleading. The excessively detailed model is unlikely to be useful or even understandable. The aim is to suppress the less necessary detail in favor of the important processes which control chemical fate.

In this chapter a number of concepts are introduced which are used to simplify and rationalize descriptions of the environment. But first, it is essential to define the system of units and dimensions which forms the foundation of all calculations.

2.2 UNITS

The introduction of the "SI" or "Système International d'Unités" or International System of Units in 1960 has greatly simplified scientific calculations. With few exceptions we will adopt the SI system. The system is particularly convenient because it is "coherent" in that the basic units combine one-to-one to give the derived units directly with no conversion factors. For example, energy (joules) is variously the product of: force (newtons) and distance (meters); or pressure (pascals) and volume (cubic meters); or power (watts) and time (seconds). Thus the foot-pound, the calorie, the liter-atmosphere and the kilowatt-hour become obsolete in favor of the single joule. Some aspects of the system are discussed below. Conversion tables from obsolete or obsolescent unit systems are available in scientific handbooks, such as the *Handbook of Chemistry and Physics* edited by Weast and published annually by CRC Press.

Length

Meter (m). This base unit is defined as the specified number of wavelengths of a krypton light emission.

Area

Square meter (m^2). Occasionally the hectare (ha) (an area 100 m square or 10^4 m^2) or the square kilometer (km^2) are used. For example, pesticide dosages to soils may be given in kg/ha.

Volume

Cubic meter (m^3). The liter (L) (0.001 m^3) is also used because of its convenience in analysis, but it should be avoided in environmental calculations.

Mass (kg)

The base unit is the kilogram (kg), but it is often more convenient to use the gram (g), especially for concentrations. For large masses the megagram (Mg) or the equivalent metric tonne (t) may be used.

Amount of Matter (mole abbreviated to mol)

This unit, which is of great importance in environmental chemistry, is really a number of constituent entities or particles such as atoms, ions, or molecules. It is the actual number of particles divided by Avogadro's number (6.0×10^{23}), which is defined as the number of atoms in 12 g of the carbon-12 isotope. When reactions occur, the amounts of substances reacting and forming are best expressed in moles rather than mass, since atoms or molecules combine in simple stoichiometric ratios. The need to involve atomic or molecular masses is thus avoided.

Molecular Mass (or Weight) (g/mol)

This is the mass of 1 mole of matter. Strictly, the correct unit is kg/mol but it is often more convenient to use g/mol which is obtained by adding the atomic masses (weights). Benzene (C_6H_6) is thus approximately 78 g/mol or 0.078 kg/mol.

Time (s or h)

The standard unit of a second (s) is inconveniently short when considering environmental processes such as flow in large lakes in which residence times may be many years. The use of hours (h), days, and even years is thus preferable. In this work we generally use hours as a compromise.

Concentration (mol/m^3 or g/m^3)

The preferred unit for environmental calculations is the mole per cubic meter (mol/m^3) or the gram per cubic meter (g/m^3). Most analytical data are reported in amount or mass per liter (L), because a liter is a convenient volume for the analytical chemist to handle and measure. Complications arise if the liter is used in environmental calculations because it is not coherent with area or length. The common mg/L, which is often ambiguously termed the "part per million," is equivalent to g/m^3. In some circumstances the use of mass fraction, volume fraction, or mole fraction as concentrations are desirable.

It is acceptable and usual to report concentrations in units such as mol/L or mg/L, but prior to any calculation they should be converted to a coherent unit of amount of substance per cubic meter.

Concentrations such as parts per thousand (ppt), per million (ppm), per billion (ppb), and per trillion (also ppt) should not be used. There can be confusion between parts per thousand and per trillion. The billion is 10^9 in North America and 10^{12} in Europe. The air ppm is usually on a volume/volume basis, whereas the water ppm is usually on a mass/volume basis.

Density (kg/m^3)

This has identical units to mass concentration but the use of kg/m^3 is preferred, water having a density of 1000 kg/m^3, and air a density of approximately 1.2 kg/m^3. Specific gravity is the dimensionless ratio of a density of a phase to the density of water (or some other substance) at a defined temperature.

Force (N)

The newton is the force which causes a mass of 1 kg to accelerate at 1 m/s^2. It is 10^5 dynes and is approximately the gravitational force operating on a mass of 102 g at the Earth's surface.

Pressure (Pa)

The pascal or newton per square meter (N/m^2) is inconveniently small since it corresponds to only 102 grams force over one square meter, but it is the standard unit, and it is used here. The atmosphere (atm) is 101325 Pa or 101.325 kPa. The torr or mm of mercury (mmHg) is 133 Pa, and although still widely used, should be regarded as obsolescent.

Energy (J)

The joule, which is one N.m or $Pa.m^3$, is also a small quantity. It replaces the obsolete units of calorie (which is 4.184 J), and BTU (1055 J).

Temperature (K)

The kelvin is preferred, although environmental temperatures are normally expressed in degrees Celsius (not centigrade) °C where 0°C is 273.16 K. There is no degree sign prior to K.

Frequency (Hz)

The hertz is one event per second (s^{-1}). It is used in descriptions of acoustic and electromagnetic waves, stirring, and in nuclear disintegration processes where the quantity of a radioactive material may be described in becquerels (Bq), where 1 Bq corresponds to the amount which has a disintegration rate of 1 Hz. The curie (Ci) which corresponds to 3.7×10^{10} disintegrations per second (and thus 3.7×10^{10} Bq) was formerly used.

Gas Constant (R)

This constant which derives from the ideal gas law is 8.314 J/mol.K or $Pa.m^3/mol.K$. An advantage of the SI system is that values of R in diverse units such as cal/mol.K and $cm^3.atm/mol.K$ become obsolete and a single, universal value now applies.

Prefixes

The prefixes used are listed in Table 2.1. Note that these prefixes precede the unit. It is not advisable to include more than one prefix in a unit, e.g., ng/mg, although mg/kg may be acceptable because the base unit of mass is the kg; however, the equivalent μg/g is clearer. The use of expressions such as an aerial pesticide spray rate of 200 g/km^2 can be ambiguous since a k(m^2) is not equal to a (km)^2. The former style is not permissible. Expressing the rate as 2 g/ha or 0.2 mg/m^2 removes all ambiguity.

The prefixes deka, hecto, deci, and centi are restricted to lengths, areas, and volumes. A common (and disastrous) mistake is to confuse milli, micro, and nano.

We use the convention J/mol.K meaning J mol^{-1} K^{-1}. Strictly, J/(mol.K) is correct but in the interests of brevity the parentheses are omitted.

Table 2.1. SI Prefixes

Factor	Prefix	Factor	Prefix
10^1	deka da	10^{-1}	deci d
10^2	hecto h	10^{-2}	centi c
10^3	kilo k	10^{-3}	milli m
10^6	mega M	10^{-6}	micro μ
10^9	giga G	10^{-9}	nano n
10^{12}	tera T	10^{-12}	pico p
10^{15}	peta P	10^{-15}	femto f
10^{18}	exa E	10^{-18}	atto a

Dimensional Consistency

When assembling quantities in expressions or equations it is critically important that the dimensions be correct and consistent. It is always advisable to write down the units on each side of the equation, cancel where appropriate, and check that terms which add or subtract have identical units. For example, an inflow or reaction rate of a chemical to a lake may be expressed as:

An inflow rate: (water flow rate G m^3/h) $\times$ (concentration C g/m^3) = GC g/h
A reaction rate: (volume V m^3) $\times$ (rate constant k h^{-1}) $\times$ (concentration C mol/m^3) = VkC mol/h.

Obviously, it is erroneous to express the above concentration in mol/L, or the volume in cm^3. When checking units it may be necessary to allow for changes in prefixes (e.g., kg to g), and for unit conversions (e.g., h to s).

Logarithms

The preferred logarithmic quantity is the natural logarithm to the base e or 2.7183, designated as ln. Base 10 logarithms are still used for certain quantities such as the octanol-water partition coefficient and for plotting on log-log or log-linear graph paper. The natural antilog or exponential of x is written either e^x or $\exp(x)$. The base 10 log of a quantity is the natural log divided by 2.303 or ln10.

2.3   THE ENVIRONMENT AS COMPARTMENTS

It is useful to view the environment as consisting of a number of *phases* or *compartments* which are in contact. Examples are the atmosphere, terrestrial soil, a lake, the bottom sediment under the lake, suspended sediment in the lake, and biota in soil or water. The phase may be continuous (e.g., water) or consist of a number of particles which are not in contact, but which are similar in properties and all reside in one phase (e.g., atmospheric dust, or biota such as fish in water). In some cases the phases may be similar chemically but different physically, e.g., the troposphere or lower atmosphere, and the stratosphere or upper atmosphere. It may be convenient to lump all biota together as one phase or consider them as two or more classes, each as a separate phase. Some compartments are in contact; thus a chemical may migrate between them (e.g., air and water), while others are not in contact and thus migration is impossible (e.g., air and bottom sediment). Some phases are accessible in a short time to migrating chemicals (e.g., surface waters), but others are accessible only slowly (e.g., deep lake or ocean waters), or effectively not at all (e.g., deep soil or rock).

In Chapter 4 we discuss these compartments in detail, and suggest typical volumes and properties.

Homogeneity and Heterogeneity

A key concept is that of phase homogeneity and heterogeneity. Well-mixed phases such as shallow pond waters tend to be homogeneous, and internal variations or gradients in concentration or temperature are negligible. Poorly mixed phases such as soils and bottom sediments are usually heterogeneous and concentrations vary with depth. Heterogeneous concentrations are difficult to describe mathematically; thus there is a compelling incentive to assume homogeneity whenever possible. Therefore, a sediment in which a chemical is present at a concentration of 1 g/m^3 at the surface, dropping linearly to zero at a depth of 10 cm, can be described approximately as a well mixed phase with a concentration of 1 g/m^3 and 5 cm deep, or 0.5 g/m^3 and 10 cm deep. In all three cases the amount of chemical present is the same, namely 0.05 g per square meter of sediment horizontal area.

Even if a phase is not homogeneous, it may be nearly homogeneous in one or two of the three dimensions. For example, lakes may be well mixed horizontally but not vertically, thus it is possible to describe concentrations as varying only in one dimension (the vertical). A wide, shallow river may be well mixed vertically, but not horizontally in the crossflow or downflow directions.

Steady and Unsteady State

If conditions change relatively little with time, there is an incentive to assume "steady state" behavior, i.e., properties are constant or independent of time. A severe mathematical penalty is paid when time dependence has to be characterized and "unsteady state" or time-varying conditions apply.

Summary

In summary, our simplest view of the environment is that of a small number of phases, each of which is homogeneous or well-mixed and unchanging with time. When this is inadequate, the number of phases may be increased, heterogeneity may be permitted in one, two, or three dimensions, and variation with time may be included. The modeler's philosophy should be to concede each increase in complexity reluctantly, and only when necessary. Each concession results in more mathematical complexity, and the need for more data in the form of kinetic or equilibrium parameters. The model becomes less understandable and thus less likely to be used, especially by others.

This is not a new idea. William of Occam expressed the same sentiment about 650 years ago when he formulated his principle of parsimony or "Occam's Razor" that

Essentia non sunt multiplicanda praeter necessitatem

which can be translated as "What can be done with fewer (assumptions) is done in vain with more" or more colloquially in this context "Don't make models more complicated than is necessary."

2.4 MASS BALANCES

When describing a volume of the environment it is obviously essential to define its limits in space. This may be simply the boundaries of water in a pond, or the air over a city to a height of 1000 m. The volume is presumably defined exactly, as are the areas in contact with adjoining phases. Having established this control "envelope," or "volume," or "parcel," we can write equations describing the two fundamental processes of change within this envelope.

Two key physical laws, those of mass and energy conservation, are accepted as axiomatic and provide the first equations. We rarely encounter situations in which nuclear processes invalidate these laws. Mass balance equations are so important as foundations of all environmental calculations that it is worthwhile to define thoroughly and unambiguously the three types which can be set up and give illustrative examples. We do not treat energy balances here, but the approach is similar.

(i) Closed System, Steady State Equations

This is the simplest class of equation. It describes how a given, constant mass of chemical will partition between various phases of fixed volume. The basic equation states that the total amount of chemical present equals the sum of the amounts in each phase, each of these amounts usually being a product of a concentration and a volume. The system is closed or "sealed" in that no entry or exit of chemical is permitted. In environmental calculations the concentrations are usually so low that the presence of the chemical does not affect the phase volumes.

Worked Example 2.1

A three phase system consists of air (100 m³), water (50 m³) and sediment (3 m³). To this is added 2 mol (M) of a hydrocarbon such as benzene. The phase volumes are not affected by this addition because the amount of hydrocarbon is small. Subscripting air, water, and sediment symbols with A, W, and S, and designating volume as V (m³) and concentration as C (mol/m³), we can write the mass balance equation

$$\text{Total amount} = \text{sum of amounts in each phase}$$
$$M = 2 = V_A C_A + V_W C_W + V_S C_S = 100\,C_A + 50\,C_W + 3\,C_S$$

To proceed further we must have information about the relationships between C_A, C_W, and C_S. This could take the form of phase equilibrium equations such as

$$C_A/C_W = 0.5 \text{ and } C_S/C_W = 100$$

These ratios are usually referred to as partition coefficients or distribution coefficients and are designated K_{AW} and K_{SW}. We discuss them in more detail in Chapter 5.

We can now eliminate C_A and C_S by substitution to give

$$2 = 100 (0.5 \ C_W) + 50 \ C_W + 3(100C_W) = 400 \ C_W$$

thus $C_W = 2/400 = 0.005 \text{ mol/m}^3$. It follows that

$$C_A = 0.5 \ C_W = 0.0025 \text{ mol/m}^3$$
$$C_S = 100 \ C_W = 0.5 \text{ mol/m}^3$$

The amounts in each phase (m mols) are the CV products as follows

$$m_W = V_W C_W = 0.250 \text{ mol. } (12.5\%)$$
$$m_A = V_A C_A = 0.250 \text{ mol. } (12.5\%)$$
$$m_S = V_S C_S = 1.500 \text{ mol. } (75\%)$$
$$\overline{\text{Total M} = 2.00 \text{ mol.}}$$

This simple algebraic procedure has established the concentrations and amounts in each phase using a closed system, steady state, mass balance equation and equilibrium relationships. The essential concept is that the total amount of chemical present must equal the sum of the individual amounts in each phase.

Example 2.2

3 mol of a pesticide of molecular mass 200 g/mol is applied to a closed system consisting of 20 m³ of water, 10 m³ of air, 1 m³ of sediment, and 0.001 m³ of biota (fish). If the concentration ratios are air/water 0.1; sediment/water 50; and biota/water 200, what are the concentrations and amounts in each phase in both gram and mole units?

Example 2.3

A circular lake of diameter 2 km and 10 m deep contains suspended solids (SS) with a volume fraction of 10^{-5}, i.e., 1 m³ of SS per 10^5 m³ water, and biota (such as fish) at a concentration of 1 mg/L. Assuming a specific gravity for biota of 1.0, a SS/water partition coefficient of 10^4, and a biota/water partition coefficient of 10^5, calculate the disposition and concentrations of 1.5 kg of a PCB in this system.

(ii) Open System, Steady State Equations

In this class of mass balance equation we introduce the possibility of the chemical flowing into and out of the system and possibly reacting, or being formed. But the conditions within the system do not change with time, i.e., its condition looks the same now as in the past and in the future. The basic equation states that the total rate of input equals the total rate of output, these rates being expressed in moles or grams per unit time. Whereas the basic unit in the closed system balance was mol or g, it is now mol/h or g/h.

Worked Example 2.4

A 10^4 m^3 well-mixed pond has a water inflow and outflow of 5 m^3/h. The inflow water contains 0.01 mol/m^3 of chemical. Chemical is also discharged directly into the lake at a rate of 0.1 mol/h. There is no reaction, volatilization, or other losses of the chemical; it all leaves in the outflow water.

(i) What is the concentration in the outflow water? We designate it as an unknown C mol/m^3.

$$\text{Total input rate} = \text{Total output rate}$$
$$5 \times 0.01 + 0.1 = 0.15 = 5C$$

thus $$C = 0.03 \text{ mol/m}^3$$

The total inflow and outflow rates are 0.15 mol/h.

(ii) If the chemical also reacts in a first order manner so that the rate is VCk mol/h, where V is the water volume, C is the chemical concentration in the well-mixed water of the pond, and k is a first order rate constant of 10^{-3} h^{-1} what will be the new concentration?

The output by reaction is VCk or $10^4 \times 10^{-3}$ C or 10 C mol/h; thus we rewrite the equation as

$$0.05 + 0.1 = 5\,C + 10\,C = 15\,C$$

thus $$C = 0.01 \text{ mol/m}^3$$

The total input of 0.15 mol/h is thus equal to the total output of 0.15 mol/h consisting of 0.05 mol/h outflow, and 0.10 mol/h reaction.

An inherent assumption is that the prevailing concentration in the pond is constant and equal to the outflow concentration. This is the "well-mixed" or "continuous stirred tank" assumption. It may not always apply, but it greatly simplifies life when it does.

The key step is to equate the sum of the input rates to the sum of the output rates, ensuring that the units are equivalent in all the terms. This often requires some unit-to-unit conversions. The unit of time must be specified carefully.

Worked Example 2.5

A lake of area (A) 10^6 m^2 and depth 10 m (volume V 10^7 m^3) receives an input of 400 mol/day of chemical in an effluent discharge. Chemical is also present in the inflow water of 10^4 m^3/day at a concentration of 0.01 mol/m^3. The chemical reacts with a first order rate constant of 10^{-3} h^{-1} and it volatilizes at a rate of $(10^{-5}C)$ mol/m^2.s, where C is the water concentration and the m^2 refers to the air-water interfacial area. The outflow is 8000 m^3/day, there being some loss of water by evaporation. Assuming that the lake water is well mixed, calculate the concentration and all the inputs and outputs in units of mol/day. Use a time unit of days.

Discharge = 400 mol/day
Inflow 10^4 × 0.01 = 100 mol/day
Total input = 500 mol/day
Reaction rate = VCk = 10^7.C.10^{-3}h^{-1}(24 h/day) = 24 × 10^4C mol/day
Volatilization rate = A(10^{-5}C) (3600 × 24 s/day) = 86.4 × 10^4C mol/day
Outflow = 8000C = 0.8 × 10^4C mol/day
Thus 500 = 24 x 10^4C + 86.4 × 10^4C + 0.8 × 10^4C = 111.2 × 10^4C
C = 4.5 × 10^{-4} mol/m^3
Reaction rate = 108 mol/day
Volatilization rate = 389 mol/day
Outflow = 3 mol/day
Total rate of loss = 500 mol/day = input rate.

Example 2.6

A building, 20 m wide, by 25 m long, by 5 m high is ventilated by introducing air from outside at a rate of 200 m^3/h. The inflow air contains CO_2 at a concentration of 0.6 g/m^3. There is also a source of CO_2 in the building of 500 g/h. What is the exit CO_2 concentration?

Example 2.7

Pesticide is applied to a 10 ha field at an average rate of 1 kg/ha every month. The soil is regarded as being 20 cm deep and well-mixed. The pesticide evaporates at a rate of 2% of the amount present per day and it degrades microbially with a rate constant of 0.05 days^{-1}. What will be the steady state concentration of pesticide (g/m^3), and in units of μg/g, assuming the soil density to be 2500 kg/m^3. What is the average standing mass of pesticide present at steady state?

In all these examples chemical is flowing or reacting, but observed conditions in the envelope do not change with time, thus steady state applies.

(iii) Unsteady State Equations

Whereas the first two types of mass balances lead to algebraic equations, unsteady state conditions give differential equations in time. The simplest method of setting up the equation is to write:

$$\text{Total input rate} - \text{Total output rate} = d(\text{Contents})/dt$$

The input and output rates should be in units such as mol/time, e.g., mol/h, the "contents" in mol, and dt the time increment (e.g., h) must be consistent with the time unit in the input and output terms. The differential equation can then be solved along with an appropriate initial or boundary condition to give an algebraic expression for concentration as a function of time. The simplest example is the first order decay equation.

Worked Example 2.8

A lake of 10^{6}m^3 is treated with 10 mol of herbicide which has a first order reaction (degradation or decay) rate constant k of 10^{-2} h^{-1}. What will be the concentration after 1 and 10 days, assuming no further input, and when will half the chemical have been degraded? Use a time unit of hours.

Here the input flow is zero. The output is only by reaction at a rate of VCk or $10^6.10^{-2}$C or 10^4C mol/h. The contents are VC or 10^6C mol. Thus,

$$0 - 10^4C = d(10^6C)/dt = 10^6 dC/dt$$

thus
$$dC/dt = -10^{-2}C \text{ mol/h}$$

This differential equation is easily solved by separating the variables C and t to give

$$dC/C = -10^{-2} dt$$

integrating gives
$$\ln C = -10^{-2}t + IC$$

where IC is an integration constant which is usually evaluated from an initial condition i.e. $C = C_o$ when $t = 0$, thus IC is $\ln C_o$ and

$$\ln(C/C_0) = -10^{-2} t$$

or
$$C = C_0 \exp(-10^{-2}t)$$

Now C_0 is $(10 \text{ mol})/10^6\text{m}^3$ or 10^{-5} mol/m³ thus

$$C = 10^{-5} \exp(-10^{-2}t)$$

After 1 day (24 h) C will be 0.79×10^{-5} mol/m³
    10 days (240 h) C will be 0.091×10^{-5} mol/m³

Half the chemical will have degraded when C/C_0 is 0.5; or $(10^{-2}t)$ is $-\ln$ 0.5 or 0.693; or t is 69.3 h. It is also possible to have inflow and outflow as well as reaction.

Worked Example 2.9

A well mixed lake of volume V 10^6 m³ containing no chemical, starts to receive an inflow of 10 m³/s water containing chemical at a concentration of

0.2 mol/m³. The chemical reacts with a first order rate constant of 10^{-2}h^{-1} and it also leaves with the outflow of 10 m³/s. What will be the concentration of chemical in the lake 1 day after the start of the input of chemical?

Input rate $= 10 \times 0.2 = 2$ mol/s (We choose a time unit of s)
Output by reaction $= (10^6 \text{ m}^3)(10^{-2}\text{h}^{-1})(1\text{h}/3600\text{s})C$ mol/m³ $= 2.77C$ mol/s
Output by flow $= 10C$ mol/s
Now, Input−Output $=$ d(Contents)/dt
$2 - 2.77C - 10C = $ d(10^6 C)/dt
or dC/$(2 - 12.77C) = 10^{-6}$dt
or $\ln(2 - 12.77C)/(-12.77) = 10^{-6}t + $ IC

When t is zero, C is zero, thus

$$\text{IC} = -\ln(2)/12.77$$

and
$$\ln[(2 - 12.77C)/2] = -12.77 \times 10^{-6} \text{ t}$$

or
$$(2 - 12.77 \text{ C}) = 2 \exp(-12.77 \times 10^{-6} \text{ t})$$

or
$$C = (2/12.77)[1 - \exp(-12.77 \times 10^{-6} \text{ t})]$$

Note that when t is zero, exp(o) is unity, and C is zero as dictated by the initial condition. When t is very large the exponential group becomes zero and C approaches (2/12.77) or 0.157 mol/m³. At such times the input of 2 mol/s is equal to the total of the output by flow of 10×0.157 or 1.57 mol/s and the output by reaction of 2.77×0.157 or 0.43 mol/s. This is a steady state condition which the lake approaches after a long period of time.

When t is 1 day or 86400s, C will be 0.104 mol/m³ or 67% of the way to its final value. C will be half way to its final value when $12.77 \times 10^{-6}t$ is 0.693 or t is 54300s or 15 hours. This time is largely controlled by the residence time of the water in the lake which is

$$(10^6 m^3)/(10 m^3/s) \text{ or } 10^5 \text{ s or } 27.7 \text{ hours}$$

Worked Example 2.10

A well mixed lake of $10^5 m^3$ is initially contaminated with chemical at a concentration of 1 mol/m³. The chemical leaves by the outflow of 0.5 m³/s and it reacts with a rate constant of 10^{-2} h⁻¹. What will be the chemical concentration after 1 and 10 days, and when will 90% of the chemical have left the lake? Use a time unit of seconds.

> Input = 0
> Output by flow = 0.5C
> Output by reaction = VCk = $10^5.C.10^{-2} h^{-1}(1/3600) = 0.278C$
> Thus $0 - 0.5C - 0.278C = 10^5 dC/dt$
> $dC/C = -0.778 \times 10^{-5} dt$
> $C = C_o \exp(-0.778 \times 10^{-5} t)$; $C_o = 1.0$
> t = 1 day = 86400s; C = 0.51 mol/m³
> t = 10 days = 864000s; C = 0.0012 mol/m³
> C = 0.1 when $0.778 \times 10^{-5} t = -\ln 0.1$ or 2.3,
>   or when t is 296000s or 3.4 days

Example 2.11

If, after the concentration of CO_2 in Example 2.6 has reached steady state, the source is reduced to 300 g/h, deduce the time course of CO_2 concentration decay to the new steady state value.

Example 2.12

Deduce the time course of pesticide concentration increase from zero to the steady state value in Example 2.7.

Example 2.13

Plot C versus time in Example 2.10, then recalculate the concentration vs time equation and plot a graph showing the effect of having a continuous emission into the lake of chemical in the inflow water of 0.5 m³/s at a concentration of 0.2 mol/m³.

Comments

These unsteady state solutions usually contain exponential terms such as exp(−kt). The term k is a characteristic rate constant with units of reciprocal time or frequency. It is thus somewhat difficult to grasp and remember. A quantity of 0.01 h⁻¹ does not convey an impression of rapidity. It is convenient to calculate its reciprocal 1/k or 100 h, which is a characteristic time. This is the time for the process to move exp(−1) or to within 37% of the final value, i.e., it is 63% completed. Those working with radioisotopes prefer to use half-lives rather than k, i.e., the time for half completion. This occurs when exp(−kt) is 0.5 or kt is ln2 or 0.693, thus the half-time t is 0.693/k, or in the above example, 69.3 hours. The other useful time is the 90% completion value, which is 2.303/k or here 230.3 h.

Mistakes are less likely to be made if rate constant data are manipulated as times rather than frequencies. Note that a rate constant of 1 day⁻¹ is 0.042h⁻¹, not 24h⁻¹; a common mistake.

If there are two first order reactions, the total rate can be calculated by adding the rate constants. This has the effect of giving a total half-time or half-life which is less than either individual half-time. It is a disastrous mistake to add half-lives.

In some cases the differential equation can become quite complex and there may be several of them applying simultaneously. Setting up these equations requires practice and care. There is a common misconception that solving the equations is the difficult task. On the contrary, it is setting them up which is most difficult and requires most skill. If the equation is difficult to solve, tables of integrals can be consulted, or an obliging mathematician can be sought. For many differential equations an analytical solution is not available and numerical methods must be used to generate a solution.

It is usually best to define the mass balance envelope as being fixed in space. When there is appreciable flow through the envelope it may be better to define the envelope as being around a certain amount of material and allow that envelope of material to change position. This "fix the material then follow it" envelope is often applied to rivers when we wish to examine the changing condition of a volume of water as it flows downstream and undergoes various reactions. It is also applied to "parcels" of air emitted from a stack and subject to wind drift. Both systems give the same results, but it may be easier to write the equations in one system than the other.

2.5   STEADY STATE AND EQUILIBRIUM

In the previous section we introduced the concept of "steady state" as implying unchanging with time, i.e., all time derivatives are zero. There can be confusion between this concept and that of "equilibrium" which can also be regarded as a situation in which no change occurs with time. The difference is important and is best illustrated by an example.

Consider the vessel in Figure 2.1A which contains 100 m³ of water and 100 m³ of air. If this is allowed to stand at constant conditions for a long time the oxygen present will equilibrate between the water and the air and reach unchanging but different concentrations, possibly 0.4 mol/m³ in the water and 8 mol/m³ in the air, i.e., a factor of 20 difference in favor of the air, with a total of 840 mol oxygen present (40 in the water and 800 in the air). In this condition the system is at equilibrium and at steady state. If half the total oxygen (420 mol) was somehow removed, but only from the air, and the volumes remained constant, the concentrations would adjust (some oxygen transferring from water to air) to give a new equilibrium (and steady state) of 0.2 mol/m³ in the water and 4 mol/m³ in the air, again a factor of 20 difference. This factor is a partition coefficient or distribution coefficient, or as is discussed later, a form of Henry's law constant. During the adjustment period (for example, immediately after removal when the air is at 3.8 mol/m³ and the water 0.4 mol/m³), the concentrations are not at a ratio of 20, the conditions are nonequilibrium, and since the concentrations are changing with time, they are also unsteady state in nature.

This correspondence between equilibrium and steady state does not, however, necessarily apply when flow conditions prevail.

It is possible for air and water to flow into and out of the tank at constant rates as shown in Figure 2.1B, and for equilibrium and steady state to be maintained, since the inflow (and outflow) concentrations are at a ratio of 20. It is even possible for near-equilibrium to apply in the vessel even if the inflow concentrations are not in equilibrium, since oxygen transfer between air and water may be very rapid. Figure 2.1B is thus a flow, equilibrium, steady state condition, whereas Figure 2.1A is a nonflow, equilibrium, steady state condition.

In Figure 2.1C there is a deficiency of oxygen in the inflow water and although in the time available some oxygen transfers from air to water, equilibrium is not reached but steady state applies because all concentrations are constant with time. This is a flow, nonequilibrium, steady state condition in which the continuous flow causes a constant displacement from equilibrium.

In Figure 2.1D the inflow water and/or air concentration or rates change with time, but there is sufficient time for the air and water to reach equilibrium in the vessel; thus equilibrium applies (the concentration ratio is always

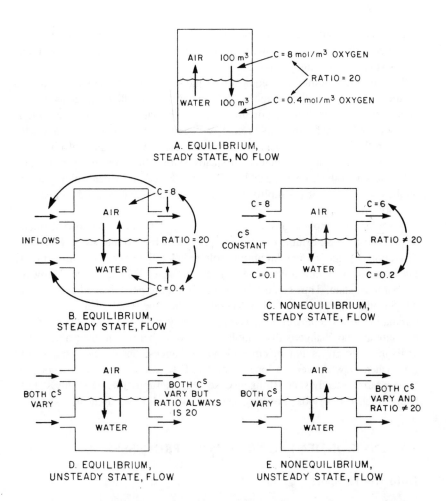

Figure 2.1. **Illustration of the difference between equilibrium and steady state conditions. Equilibrium implies that the oxygen concentrations in the air and water achieve a ratio or partition coefficient of 20. Steady state implies unchanging with time, even if flow occurs and regardless of whether equilibrium applies or not.**

20), but the system is unsteady state. Similar behavior could occur if the tank temperature or the water level was changing with time. This is a flow, equilibrium, unsteady state condition.

Finally, in Figure 2.1E the concentrations change with time and they are not in equilibrium; thus a flow, nonequilibrium, unsteady state condition applies which is obviously quite complex.

The important point is that equilibrium and steady state are not synonymous; neither, either, or both can apply. Equilibrium implies that phases have concentrations (or temperatures or pressures) such that they experience no tendency for net transfer of mass or energy. Steady state implies constancy with time. In the real environment we observe a complex assembly of phases in which some are (approximately) at steady state, others in equilibrium, and still others at both steady state and equilibrium. By carefully determining which applies we can greatly simplify the modeling mathematics used to describe the environment.

A couple of complications are worthy of note. Chemical reactions also tend to proceed to equilibrium, but may be prevented from doing so by kinetic or activation considerations. An unlit candle seems to be in equilibrium with air, but in reality it is in a metastable equilibrium state. If lit, it proceeds toward a "burned" state. Thus some reaction equilibria are not achieved easily, or at all. Second, "steady state" also depends on the time frame of interest. Blood circulation in a sleeping child is nearly in steady state; the flowrates are fairly constant and no change is discernible over several hours. But over a period of years the child grows and the circulation rate changes; thus it is not a true steady state when viewed over the long term. The first case is really a "pseudo" or "short term" steady state. In many cases it is useful to assume steady state to apply for short periods, knowing that it is not valid over long periods.

2.6   ENVIRONMENTAL TRANSPORT PROCESSES

Diffusive transport

In the air-water example it was argued that equilibrium occurs when the ratio of the air and water concentration was 20. Thus if the water concentration was 0.5 mol/m^3, equilibrium occurs when the air concentration is 10 mol/m^3. If the air concentration changes to 14 mol/m^3 we expect oxygen to transfer by diffusion from air to water, thus the air concentration falls and water concentration rises and a new equilibrium is reached at, say, 0.6 and 12 mol/m^3. This is easily calculated if the total amount of chemical is known, as was illustrated in Example 2.1.

Conversely, if the initial air concentration was reduced to 2 mol/m^3, the concentrations could adjust to C_W of 0.2 and C_A of 4. These changes are shown graphically as A to B, and C to D in Figure 2.2, in which the area

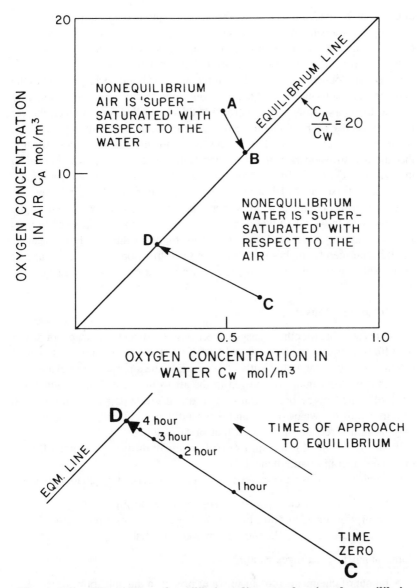

Figure 2.2.   Illustration of equilibrium diagram showing the equilibrium line with nonequilibrium regions on either side. A system initially at condition A will migrate to B, and one at C to D. The times of diffusive approach to equilibrium are shown; the concentration difference halving each hour.

of the graph is divided by the "equilibrium line" into two nonequilibrium areas. If two phases are in contact, the point corresponding to the two concentrations will always tend to migrate to the line. It can be shown that the migration route is constrained by the mass balance equation.

If we could observe and time the migration of the point corresponding to the two concentrations along this line, we would find that the point moves rapidly at first, then more slowly as it approaches the equilibrium line. In fact, the velocity is usually proportional to the distance from the line, halving over each constant time increment as shown. The reason for this (as we discuss later) is that the net diffusion rate of the oxygen from air to water is proportional to the extent of departure from equilibrium. In effect, this is a statement arising from Fick's First Law of Diffusion, which is more fully discussed in Chapter 7.

A worrisome dilemma is: how does the oxygen in the water know the concentration in the air so that it can decide to start or stop diffusing? In fact, it does not know, or care. It diffuses regardless of the condition at the destination. Equilibrium merely implies that there is no *net* diffusion, the water to air and air to water diffusion rates being equal in rate and opposite in direction.

Nondiffusive Transport

But it is possible for other transport mechanisms to occur which are not driven by diffusion. For example, we could take 1 m³ of the water with its associated 0.5 mol of oxygen and physically convey it into the air, force it to evaporate, thus causing the amount of oxygen in the air to increase. This nondiffusive or "bulk" or "piggy back" transfer occurs at a rate dependent on the rate of removal of the water phase, and is not influenced by diffusion. Indeed, it may be in a direction opposite to that of diffusion.

In the environment it transpires there are many diffusive and nondiffusive processes operating simultaneously.

Examples of diffusive transfer processes experienced by chemicals are:

- volatilization of chemical from soil or water to air
- absorption or adsorption of chemical by sediments from water
- diffusive uptake of chemical from water by fish.

Some nondiffusive processes are:

- fallout of chemical from air to water or soil in dustfall, rain, or snow
- deposition and resuspension of sediment particles and chemical
- ingestion and egestion of food and its associated chemical by biota.

The mathematical expressions for these rates are quite different. For diffusion, the rate of transfer or "flux" is written as the product of the departure from equilibrium and a kinetic quantity, and it thus becomes zero when the phases reach equilibrium. We examine these diffusive processes in Chapter 7. For nondiffusive processes, the flux is simply the product of the volume of the phase transferred per unit time (e.g., quantity of sediment or rain) and the concentration. We treat nondiffusive processes in Chapter 6.

We use the word "flux" as synonymous with transport rate with units such as mol/h or g/h. Purists insist that flux should have units of mol/h.m², i.e., it is area-specific. We will be somewhat sloppy and apply it to both. It is erroneous to use the term "flux rate" since flux, like speed or velocity, already contains the "per time" term. We would never use the term "speed rate."

2.7 FLOW AND RESIDENCE TIMES AND PERSISTENCE

In some environments such as lakes it is convenient to define a residence time or detention time. If a pond has a volume of 1000 m³ and experiences inflow and outflow of 2 m³/h it is apparent that on the average the water spends 500 h (i.e., 1000 m³/2 m³/h) in the lake. This residence or detention time may not bear much relationship to the actual time which a particular parcel of water spends in the pond, since some water may bypass most of the pond and reside for only a short time, and some may be trapped for years. The quantity is useful, however, because it provides an immediate insight into the time required to flush out the contents. Obviously a large lake with a long residence time will be very slow to recover from contamination. Comparison of the residence time with a chemical reaction time (e.g., a half-life) indicates whether a chemical is likely to be removed from a lake predominantly by flow or by reaction.

If a well-mixed lake has a volume V m³ and inflow and outflow rates G m³/h, then the flow residence time t_F is V/G(h). If it is contaminated by a nonreacting (conservative) chemical at a concentration C_O mol/m³ at zero time and there is no new emission, a mass balance gives, as was shown earlier

$$C = C_0 \exp(-Gt/V) = C_0 \exp(-t/t_F)$$

The residence time is thus the reciprocal of a rate constant k_F which has units of h⁻¹. The half-time for recovery occurs when t/t_F or $k_F t$ is ln 2 or 0.693, i.e., when t is $0.693 t_F$ or $0.693/k_F$.

If the chemical also undergoes a reaction with a rate constant k_R h^{-1} it can be shown that

$$C = C_0 \exp-[(k_F + k_R)t] = C_0 \exp(-k_T t)$$

thus the larger (faster) rate constant dominates. The characteristic times t_F and t_R (i.e., $1/k_F$ and $1/k_R$) combine as reciprocals to give the total time t_T, as do electrical resistances in parallel, i.e.,

$$1/t_F + 1/t_R = 1/t_T$$

thus the smaller (shorter) t dominates. The term t_R can be viewed as a reaction persistence. Characteristic times such as t_R and t_F are conceptually easy to grasp, and are very convenient quantities to deduce when interpreting the relative importance of environmental processes. For example, if t_F is 30 years and t_R is 3 years, t_T is 2.73 years; thus reaction dominates the chemical's fate. Ten out of every 11 molecules react and only one leaves the lake by flow.

Later, we will frequently use the word ''advection'' as meaning the transport of a chemical from a region by bulk flow. For example, chemical in solution is advected down rivers, or gaseous chemical is advected with the wind. The related term ''convection'' is usually applied to transport of fluid induced by density differences resulting from heating. Although widely used in environmental circles, the word ''advection'' is absent from many dictionaries.

2.8   REAL AND EVALUATIVE ENVIRONMENTS

The environmental scientist who is struggling to describe the behavior of a pesticide in a system such as a pond soon discovers that real ponds are very complex. Considerable effort is required to measure, analyze, and describe the pond with the result that little energy (or research money) remains with which to describe the behavior of the pesticide. This is an annoying problem because it diverts attention from the pesticide (which is important) to the condition of the pond (which may be relatively unimportant). A related problem arises when a new chemical is being considered. Into which pond should it be placed (hypothetically) for evaluation? A significant advance in environmental science was made in 1978 when Baughman and Lassiter (1978) proposed that chemicals may be assessed in ''evaluative environments'' which have fictitious, but realistic, properties such as volume, composition, and temperature. Evaluative environments can be decreed to consist of a few homogeneous phases of specified dimensions with constant temperature and composition.

Essentially, the environmental scientist designs a "world" to desired specifications, then explores mathematically the likely behavior of chemicals in that world. No claim is made that the evaluative world is identical to any real environment, although broad similarities in chemical behavior are expected. Evaluative aquatic environments were used to develop the celebrated and pioneering U.S. Environmental Protection Agency (EPA) EXAMS model of chemical fate in rivers and lakes.

There are good precedents for this approach. Carnot devised an evaluative steam engine, now termed the Carnot cycle, which leads to a satisfying explanation of entropy and the Second Law of Thermodynamics. The kinetic theory of gases uses an evaluative assumption of gas molecule behavior.

The principal advantage of evaluative environments is that they act as an intellectual stepping-stone when tackling the difficult task of describing both chemical behavior and an environment. The task is simplified by sidestepping the effort needed to describe a real environment. The disadvantage is that results of evaluative environment calculations cannot be validated directly; they are thus suspect, and possibly quite wrong. Some validation can be sought by making the evaluative environment similar to a simple real environment such as a small pond, or a laboratory microcosm.

Later, we construct evaluative environments or "units worlds" and use them to explore the likely behavior of chemicals. In doing so we use equations which can be validated using real environments. A somewhat different assembly of equations proves to be convenient for real environments, but the underlying principles are the same.

2.9 SUMMARY

In this chapter we have introduced the system of units and dimensions. A view of the environment has been presented as an assembly of phases or compartments which are (we hope) mostly homogeneous rather than heterogeneous in properties, and which can vary greatly in volume and composition. We can define these phases, or parts of them, as "envelopes" about which we can write mass balance equations. Steady state conditions will yield algebraic equations, and unsteady state conditions will yield differential equations. These equations may contain terms for flow (diffusive and nondiffusive) of chemical between phases, and for reaction or formation of a chemical. We have discriminated between equilibrium and steady state, and introduced the concepts of residence time and persistence. Finally, the use of both real and evaluative environments has been suggested.

Having established these basic concepts, or working tools, our next task is to develop in subsequent chapters the capability of quantifying the rates of the various flow, transport, and reaction processes to which chemicals present in the environment are subject.

It is first useful to digress briefly to examine some chemicals which are of concern, which we do in Chapter 3. We then examine the nature of environmental compartments in Chapter 4. The first fundamental mathematical task (which we address in Chapter 5) is to define the conditions of chemical equilibrium between phases. If equilibrium applies, no net diffusive flow will occur. If equilibrium does not apply we can at least predict in which direction diffusive flow will occur. Calculating how *fast* it occurs requires a knowledge of (a) the extent of departure from equilibrium and (b) kinetic or transport rate information, the latter being postponed until Chapter 7. Chemical removal processes such as reaction and advection are examined in Chapter 6. Finally, Chapter 8 illustrates and applies these capabilities in complete mathematical models of real environments.

3 ENVIRONMENTAL CHEMICALS AND THEIR PROPERTIES

3.1 INTRODUCTION

There is a formidable literature on the nature and properties of chemicals of environmental concern. Several handbooks list relevant physical-chemical and toxicological properties. Especially extensive are compilations on pesticides, on chemicals of potential occupational exposure, and carcinogens. Government agencies such as the United States Environmental Protection Agency (U.S. EPA), scientific groups such as the U.S. National Academy of Sciences, as well as individual authors, have published many reports and books on specific chemicals or classes of chemicals. Conferences are regularly held and the proceedings published on chemicals such as the "dioxins." Computer-accessible databases are now widely available for consultation. A useful starting point is the handbook by Verschueren (1983) which gives environmentally relevant data, and the series by Howard (1989), which gives more detailed chemical-by-chemical assessments. In the space available here we can only present a few key concepts, this chapter being based largely on material written by the author for the Canadian Environmental Advisory Council (CEAC, 1988).

Most of the chemicals that we treat in this book are organic, but we also discuss metals, some organo-metallic chemicals, gases such as oxygen and methane, inorganic compounds, and ions containing elements such as phosphorus and arsenic. Metals and inorganic compounds tend to require individual treatment because they tend to possess a unique set of properties. Organic compounds, on the other hand, tend to fall into certain well-defined classes. We are thus able to estimate the properties and behavior of one organic chemical from those of other, somewhat similar, or homologous chemicals.

It is believed that over 60,000 chemicals are used in commerce. The number of chemicals of possible concern runs to about 1500. There are now numerous lists of "priority" chemicals, but there is considerable variation between lists. It is not possible, or even useful, to specify an exact number of chemicals. Some inorganic chemicals ionize in contact with water, and thus lose their initial identity. Some lists name PCBs (polychlorinated biphenyls) as one chemical, others as six groups of chemicals, whereas in reality, the PCBs consist of 209 possible individual chemicals. It is only recently that analytical chemists have identified all 209 congeners. Many chemicals, such as surfactants and solvents, are complex mixtures which are difficult to identify and analyze. One designation, such as "naphtha," may represent 1000 chemicals.

For legislative purposes, most jurisdictions have compiled lists of chemicals which are, or may be, encountered in the environment, and from these "raw" lists of chemicals of potential concern, have established smaller lists of "priority" chemicals. These are chemicals which are usually observed in the environment, are known to have the potential to cause adverse ecological/biological effects, and are thus believed to be worthy of control and regulation. In practice a chemical which fails to reach the "priority" list is usually ignored and receives no priority, not less priority.

The first and most celebrated list was the U.S. Environmental Protection Agency's Priority Pollutants list, or EPA list, which originally included 126 chemicals or groups. This list was established as a result of a legal agreement between the U.S. EPA and the Natural Resource Defense Council in 1976. International organizations, such as OECD, have established somewhat similar lists, as has the Canadian federal government under the Canadian Environmental Protection Act. The State of Michigan has a Critical Materials Register, which is an expansion of the EPA list. The International Agency for Research on Cancer (IARC) produces a list of chemicals suspected to be carcinogenic. The United Nations Environment Programme has an International Register of Potentially Toxic Chemicals (IRPTC) and issues a regular bulletin describing international developments in toxic chemical assessment.

Agencies concerned with particular ecosystems, such as the North American Great Lakes, have established lists of chemicals found throughout that ecosystem, an example being an International Joint Commission "Inventory of Chemicals Identified in the Great Lakes System." There are also lists of chemicals found in specific parts of ecosystems; for example, the Niagara River Toxic Chemical list. In Canada, the most comprehensive and thoughtfully prepared list is the Province of Ontario's Effluent Monitoring Priority Pollutants List, which started as some 180 chemicals or groups which were identified as being of particular concern to that Province. The list, like others, continues to grow as new chemicals are discovered in the environment by enthusiastic analytical chemists.

Table 3.1 lists about 200 chemicals by class and includes most of the chemicals of current concern.

Table 3.1.   List of Chemicals Commonly Found on Priority Chemical Lists

Volatile halogenated hydrocarbons
Chloromethane
Methylene chloride
Chloroform
Carbon tetrachloride

Chloroethane
1,1-Dichloroethane
1,2-Dichloroethane
1,1-Dichloroethylene
trans-1,2-Dichloroethylene
Vinyl chloride

1,1,1-Trichloroethane
1,1,2-Trichloroethane
Trichloroethylene

1,1,2,2-Tetrachloroethane
Tetrachloroethylene
Hexachloroethane
1,2-Dichloropropane
1,3-Dichloropropane
cis-1,3-Dichloropropylene
trans-1,3-Dichloropropylene

Hexachlorobutadiene

Hexachlorocyclopentadiene
Chloroprene
Bromomethane
Bromoform

Ethylenedibromide

Chlorodibromomethane
Dichlorobromomethane
Dichlorodibromomethane
Freons (chlorofluoro-hydrocarbons)
Dichlorodifluoromethane
Trichlorofluoromethane

Alkenes
1,3-Butadiene

Monoaromatic hydrocarbons
Benzene
Toluene
o-Xylene
m-Xylene
p-Xylene
Ethylbenzene
Styrene

Polycyclic aromatic hydrocarbons (PAHs)
Naphthalene
1-Methylnaphthalene
2-Methylnaphthalene
Trimethylnaphthalene
Biphenyl

(continued on next page)

Table 3.1. (Continued)

Acenaphthene
Acenaphthylene
Fluorene
Anthracene
Fluoranthene
Phenanthrene
Pyrene
Chrysene
Benzo(a)anthracene
Dibenz(a,h)anthracene
Benzo(b)fluoranthene
Benzo(k)fluoranthene
Benzo(a)pyrene
Perylene
Benzo(ghi)perylene
Indeno(1,2,3)pyrene

Halogenated aromatics
Chlorobenzene
1,2-Dichlorobenzene
1,3-Dichlorobenzene
1,4-Dichlorobenzene
1,2,3-Trichlorobenzene
1,2,4-Trichlorobenzene
1,2,3,4-Tetrachlorobenzene
1,2,3,5-Tetrachlorobenzene
1,2,4,5-Tetrachlorobenzene
Pentachlorobenzene
Hexachlorobenzene
2,4,5-Trichlorotoluene
Octachlorostyrene
1-Chloronaphthalene
2-Chloronaphthalene

**Biphenyls and halogenated
  biphenyls**
Biphenyl

Polychlorinated biphenyls (PCBs)
Polybrominated biphenyls (PBBs)
Aroclor Mixtures (PCBs)
Aroclor 1016
Aroclor 1221
Aroclor 1232
Aroclor 1242
Aroclor 1248
Aroclor 1254
Aroclor 1260

Chlorinated dibenzo-p-dioxins
2,3,7,8-Tetrachlorodibenzo-p-dioxin
Tetrachlorinated dibenzo-p-dioxins
Pentachlorinated dibenzo-p-dioxins
Hexachlorinated dibenzo-p-dioxins
Heptachlorinated dibenzo-p-dioxins
Octachlorodibenzo-p-dioxin
Brominated dibenzo-p-dioxins

Chlorinated dibenzofurans
Tetrachlorinated dibenzofurans
Pentachlorinated dibenzofurans
Hexachlorinated dibenzofurans
Heptachlorinated dibenzofurans
Octachlorodibenzofuran

Alcohols and phenols
Benzyl alcohol
Phenol
o-Cresol
m-Cresol
p-Cresol
2-Hydroxybiphenyl
4-Hydroxybiphenyl
Eugenol

(continued on next page)

Table 3.1   (Contin'

Halogenated phenols
2-Chlorophenol
2,4-Dichlorophenol
2,6-Dichlorophenol
2,3,4-Trichlorophenol
2,3,5-Trichlorophenol
2,4,5-Trichlorophenol
2,4,6-Trichlorophenol
2,3,4,5-Tetrachlorophenol
2,3,4,6-Tetrachlorophenol
2,3,5,6-Tetrachlorophenol
Pentachlorophenol

4-Chloro-3-methylphenol
2,4-Dimethylphenol
2,6-Di-t-Butyl-4-Methylphenol
Tetrachloroguaiacol

**Nitrophenols, nitrotoluenes
and related compounds**
2-Nitrophenol
4-Nitrophenol
2,4-Dinitrophenol
4,6-Dinitro-o-cresol

Nitrobenzene
2,4-Dinitrotoluene
2,6-Dinitrotoluene
1-Nitronaphthalene
2-Nitronaphthalene
5-Nitroacenaphthene

Nitrogen and sulfur compounds
N-Nitrosodimethylamine
N-Nitrosodiethylamine
N-Nitrosodiphenylamine
N-Nitrosodi-n-propylamine

Dip⊦
Indol⸝
4-Aminoa⸝
Acrylonitrile
Benzidine
3,3-Dichlorobenzidine
Benzeneacetonitrile
Aniline
Acridine
Ethylenethiourea
Hydrazine
2-Mercaptobenzothiazole
Morpholine

Acids
Abietic Acid
Dehydroabietic Acid
Chlorodehydroabietic Acid
Oleic Acid
Pimaric Acid

**Ethers, ketones, aldehydes,
and related compounds**
Diphenylether
4-Chlorophenyl-phenylether
bis(2-chloromethyl)ether
bis(2-chloroethyl)ether
bis(2-chloroisopropyl)ether
4-Bromophenyl-phenylether
Diphenyl ether
bis(2-chloroethoxy)methane
Formaldehyde
Benzaldehyde
Butanal
Methylethylketone

(continued on next page)

Table 3.1. (Continued)

thers
,4-Dioxane
Dimethyl disulfide

Phthalate esters
Dimethylphthalate
Diethylphthalate
Di-n-butylphthalate
Di-n-octylphthalate
Di(2-ethylhexyl) phthalate
Benzylbutylphthalate

Pesticides
Acrolein
Aldicarb
Aldrin
Alachlor
Atrazine
α-Endosulfan
β-Endosulfan
Endosulfan sulfate
α-BHC
β-BHC
δ-BHC
γ-BHC (Lindane)
Chlordane
Chlorpyrifos
Dicamba
Dieldrin
4,4'-DDE
4,4'-DDD
4,4'-DDT
Endrin
Fenitrothion
Heptachlor
Heptachlor epoxide
Isophorone
Malathion

Parathion
Methylparathion
Methoxychlor
Mirex
Toxaphene

Metals and inorganic compounds
Aluminum
Antimony
Beryllium
Cadmium
Chromium (especially hexavalent)
Cobalt
Copper
Lead (especially organic)
Mercury (especially organic)
Molybdenum
Nickel
Silver
Thallium
Tin (especially organic)
Vanadium
Zinc
Cyanides
Arsenic
Phosphorus
Selenium
Chloramines
Chlorine
Ozone
Hydrogen sulfide
Ammonia
Carbon monoxide
Sulfur dioxide

Asbestos

Radionuclides
Radon and its progeny

3.2 SELECTING PRIORITY CHEMICALS

It is a challenging task to select from "raw lists" of chemicals a smaller manageable number of "priority" chemicals. Such chemicals receive intense scrutiny, analytical protocols are developed, properties and toxicity are measured, and reviews conducted of sources, fate, and effects. This selection contains an element of judgment and is approached by different groups in different ways. A common thread to many of these selection processes is the consideration of five factors: quantity, persistence, bioaccumulation, toxicity, and a miscellaneous group of "other adverse effects."

Quantity

The first factor is the quantity produced, used, formed or transported, including consideration of the fraction of the chemical which may be discharged to the environment during use. Some chemicals, such as benzene, are used in very large quantities in fuels, but only a small fraction, (possibly a percent or less), is emitted into the environment through incomplete combustion or leakage during storage. Other chemicals, such as pesticides, are used in much smaller quantities, but are discharged totally and directly into the environment, i.e., 100% is emitted. On the other extreme are chemical intermediates which may be produced in large quantities but are emitted in only miniscule amounts, except during an industrial accident. It is difficult to compare the amounts emitted from these various categories because they are highly variable and episodic. It is essential, however, to consider this factor because many toxic chemicals have no real adverse impact because they enter the environment in negligible quantities.

Persistence

The second factor is the chemical's environmental persistence, which may also be expressed as a lifetime, half-life, or residence time. Some chemicals, such as DDT or the PCBs, may persist in the environment for several years by virtue of their resistance to transformation or degrading processes of biological and physical origin. They have the opportunity to migrate widely throughout the environment, and may reach vulnerable organisms. Their persistence results in the possibility of establishing relatively high concentrations. This arises because the amount in the environment (kilograms) can be expressed as the product of the emission rate into the environment (kilograms per year) and the residence time of the chemical in the environment (years).

This is the same equation which controls a human population. For example, the number of Canadians (about 25 million) is determined by the product of the rate at which Canadians are born (about 1/3 million per year), and the lifetime of Canadians (about 75 years). If Canadians lived for only 30 years, the population would drop to 10 million.

Intuitively, the amount (and hence the concentration) of a chemical in the environment must control the effect of that chemical on ecosystems because toxic effects generally respond to concentrations. Unfortunately, it is difficult to estimate the environmental persistence of a chemical. This is because the rate at which chemicals degrade depends on which environmental media they reside in, on temperature (which varies diurnally and seasonally), on incidence of sunlight which varies similarly, on the nature and number of degrading microorganisms which may be present, and on other factors such as acidity and the presence of catalysts. This variable persistence contrasts with radio-isotopes, which have a half-life which is fixed and unaffected by the media in which they reside. Obviously long-lived chemicals such as PCBs are of much greater concern than those such as phenol, which may persist in the aquatic environment for only a few days as a result of susceptibility to biodegradation. Some estimate of persistence or residence time is thus necessary for priority-setting purposes. Organo-chlorine chemicals tend to be persistent and are thus frequently found on priority lists.

Bioaccumulation

The third factor is potential for bioaccumulation (i.e., uptake of chemical by organisms). This is a phenomenon, not an effect. Thus bioaccumulation *per se* is not necessarily of concern. It is of concern that bioaccumulation may cause toxicity to the affected organism, or to a predator or consumer of that organism. Historically, it was observation of pesticide bioaccumulation in birds which prompted Rachel Carson to write *Silent Spring* in 1962, thus greatly increasing public awareness of environmental contamination.

Some chemicals, notably the hydrophobic or "water-hating" organic chemicals, partition appreciably into organic media and establish high concentrations in fatty tissue. PCBs may achieve concentrations (i.e., they bioconcentrate) in fish at factors of 100,000 times the concentrations in the water in which the fish dwell. For some chemicals (notably PCBs, mercury, DDT) there is also a food chain effect. Small fish are consumed by larger fish, at higher trophic levels, and by other animals, such as gulls, otters, mink, and humans. These chemicals may be transmitted up the food chain, and in doing so may experience a further increase in concentration so that they are biomagnified.

Bioaccumulation tendency is normally estimated using an organic phase-water partition coefficient and, more specifically, the octanol-water partition coefficient. This, in turn, can be related to the solubility of the chemical in the water.

Clearly, chemicals which bioaccumulate, bioconcentrate, and biomagnify have the potential to cause unusual toxic effects, can travel unusual pathways, and can exert severe toxic effects, especially on organisms at higher trophic levels.

The importance of bioaccumulation may be illustrated by noting that in water containing 1 ng/L of PCB, the fish may contain 10^5 ng/L or equivalently 10^5 ng/kg. A human may consume 1000 L of water annually (containing 1000 ng) and 10 kg of fish (containing 10^6 ng), thus exposure is primarily from fish consumption.

Toxicity

The fourth factor is the toxicity of the chemical. The simplest manifestation of toxicity is acute toxicity. This is most easily measured as a concentration which will kill 50% of a population of an aquatic organism, such as a fish or an invertebrate (e.g., *Daphnia magna*), in a period of 24–96 hours, depending on test conditions. When the concentration which kills (or is lethal to) 50% (the LC50) is small, this corresponds to high toxicity. The toxic chemical may also be administered to laboratory animals such as mice or rats, orally or dermally. The results are then expressed as a lethal dose to 50% (LD50) in units of mg chemical/kg body weight of the animal.

More difficult, expensive, and contentious are chronic, or sub-lethal tests, which assess the susceptibility of the organism to adverse effects from dosages or concentrations of chemicals which do not cause immediate death, but may lead ultimately to death. For example, the animal may cease to feed, grow more slowly, be unable to reproduce, become more susceptible to predation, or display some abnormal behavior which ultimately affects its life span or performance. The concentrations or dosages at which these effects occur are often about 1/10th to 1/100th of those which cause acute effects. Ironically, in many cases the toxic agent is also an essential nutrient, so too much or too little may cause adverse effects.

Although most toxicology is applied to animals there is also a body of knowledge on toxicity to plants or phytotoxicity. Plants are much easier to manage, and killing them is less controversial. Tests also exist for assessing toxicity to microorganisms.

It is important to emphasize that toxicity is not a sufficient cause for concern about a chemical. Arsenic in a bottle is harmless. Disinfectants are inherently useful because they are toxic. A fundamental principle of toxicology is that the "dose makes the poison"; the extent to which the organism is injured depends on the inherent properties of the chemical AND the dose or amount which the organism experiences. It is thus misleading to classify or prioritize chemicals solely on the basis of their inherent toxicity, or on the basis of the concentrations or exposure present in the environment. Both must be considered. A major task of this book is to estimate exposure. A healthy tension often exists between toxicologists and chemists about the relative importance of toxicity and exposure, but fundamentally this argument is about as purposeful as squabbling over whether tea leaves or water are the more important constituents of tea.

Most difficult is the issue of genotoxicity, including carcinogenicity, mutagenicity, and teratogenicity. In recent years, a battery of tests has been developed in which organisms, ranging from microorganisms to mammals, are exposed to chemicals in an attempt to determine if they can influence genetic structure or cause cancer. A major difficulty is that these effects may have long latent periods, perhaps 20 to 30 years in humans. The adverse effect may be a result of a series of biochemical events in which the toxic chemical plays only one role. It is difficult to extrapolate from the results of short-term laboratory experiments to deduce reliably the presence and magnitude of hazard to humans. There may be suspicions that a chemical may be producing cancer in perhaps 0.1% of a large human population over a period of perhaps 30 years; an effect which is very difficult (or probably impossible) to detect in epidemiological studies. But this 0.1% translates into the premature death of 300 Canadians per year from such a cancer, and is cause for considerable concern. Another difficulty is that humans are voluntarily and involuntarily exposed to a large number of toxic chemicals including those derived from smoking, legal and illicit drugs, domestic and occupational exposure, as well as environmental exposure. Despite these difficulties, a considerable number of chemicals have been assessed as being carcinogenic, mutagenic, or teratogenic, and it is even possible to assign some degree of potency to each chemical. Such chemicals usually rank high in priority lists.

Other Effects

Finally, there is a variety of other adverse effects which are of concern, including:

- the ability to influence atmospheric chemistry (e.g., freons)
- alteration in pH (e.g., acid rain)
- unusual chemical properties such as chelating capacity which alters the availability of other chemicals in the environment
- interference with visibility
- odor (e.g., from organo-sulfur compounds)
- color (e.g., from dyes)
- the ability to cause foaming in rivers (e.g., detergents or surfactants).

Selection Procedures

A common selection procedure involves scoring these factors on some hazard scale, for example, 1 to 10. The factors may then be combined to give an overall factor and determine priority. This is a subjective process and becomes difficult for two major reasons.

First, chemicals which are subject to quite different patterns of use are difficult to compare. For example, chemical X may be produced in very large quantities, is emitted directly into the environment and is found in substantial concentrations in the environment, but is not believed to be particularly toxic. Examples are solvents such as trichloroethylene or plasticizers such as phthalate esters. On the other hand chemical Y may be produced in miniscule amounts but is very toxic, an example being the "dioxins." Which deserves the higher priority?

Second, it appears that the adverse effects suffered by aquatic organisms and other animals, including humans, are the result of exposure to a large number of chemicals, and not just to one or two chemicals. Thus, assessing chemicals on a case by case basis may obscure the cumulative effect of a large number of chemicals. For example, if an organism is exposed to 150 chemicals, each at a concentration which is only 1% of the level which will cause death, then death is very likely to occur, but it cannot be attributed to any one of these chemicals. It is the cumulative effect which causes death. The obvious prudent approach is to reduce exposure to all chemicals to the maximum extent possible. The issue is further complicated by the possibility that some chemicals will act synergistically, i.e., produce an effect which is greater than additive, or antagonistically, i.e., the combined effect is less than additive.

The result is that there will probably be cases in which we are unable to prove that a specific chemical causes toxicity, but in reality it does contribute to an overall toxic effect. Indeed, some believe that this situation will be the rule rather than the exception.

A compelling case can be made that the prudent course of action is for society to cast a fairly wide net of suspicion (i.e., assemble a fairly large list of chemicals), then work to elucidate sources, fate, and effects with the aim of reducing overall exposure of humans and our companion organisms, to a level where there is assurance that no significant toxic effects can exist from these chemicals. The risk from these chemicals then becomes small compared to other risks such as accidents, disease, and exposure to natural toxic substances. This may, or may not, make the risk acceptable.

3.3 KEY CHEMICAL PROPERTIES AND CLASSES

3.3.1 Introduction

Fundamental to the assessment of a chemical's fate and effects is a knowledge of its structure, and hence its molecular mass. Melting point, boiling point, density, and occasionally viscosity are also valuable.

Organic chemicals have a tendency to be "hydrophobic" or correspondingly "lipophilic." This implies that the chemical is sparingly soluble in, or "hates" water and prefers to partition into lipid or organic or fat phases. A convenient descriptor of this hydrophobic tendency is the octanol-water partition coefficient which is the ratio of the equilibrium concentration which the chemical adopts in octanol to that in water. A high value of perhaps one million, as applies to DDT, implies that the chemical will achieve a concentration in an organic medium approximately a million times that of water which is in contact with it. In reality, most organic chemicals are approximately equally soluble in lipid or fat phases but they vary greatly in their solubility in water. Thus differences in hydrophobicity are largely due to differences of behavior in, or affinity for, the water phase. This is most conveniently characterized by the solubility of the chemical in water. For example, benzene has a solubility of 1780 g/m^3. Many organic chemicals such as ethanol do not have a reported solubility in water because they are miscible with water in all proportions. A low water solubility implies high hydrophobicity and a high octanol-water partition coefficient. Before any attempt is made to characterize the behavior of an organic chemical in the environment, it is essential to know its water solubility, and its octanol-water partition coefficient.

The chemical's tendency to partition into the atmosphere is controlled by its vapor pressure, which is essentially the maximum pressure that pure chemical can exert in the atmosphere. This can be viewed as a kind of "solubility" of the chemical in the atmosphere. Indeed, if the vapor pressure P (Pa) is divided by the gas constant, temperature group RT, where R is the gas constant (8.314 $Pa.m^3/mol.K$) and T is absolute temperature (K), then the vapor pressure can be converted into a solubility C (mol/m^3). Organic chemicals vary enormously in their vapor pressure and correspondingly in their boiling point. Some such as the lower alkanes, which are present in gasoline, are very volatile, whereas others such as DDT have exceedingly low vapor pressures.

The ratio of the solubility in air to solubility in water is essentially an air-water partition coefficient or a version of the Henry's law constant. A frequent problem facing analytical chemists extracting water samples is that of separating the organic chemical from the water. A convenient way of accomplishing this is to bubble or sparge air through the water and strip the chemical from the solution. This is effective if the air-water partition coefficient exceeds about 0.01. Such chemicals are said to be "purgable" and are often referred to simply as "volatiles." In lists of organic chemicals they are often separated into a special group because of their susceptibility to this convenient analytical method. The gas stream containing the evaporated chemical can be passed through a trap which will retain the chemical. The chemical is later thermally desorbed and introduced to an analytical instrument such as a gas chromatograph.

Another major classification of organic chemicals is according to their dissociating tendencies in water solution. Some organic acids, notably the phenols, will form ionic species (phenolates) at high pH. This enables these compounds to be extracted by base from an organic extract and thus analyzed separately. This leaves the basic and neutral chemicals for subsequent analysis.

We examine in the following sections a number of classes of compounds which are of concern in this book. In doing so we tabulate some properties of molecular weight, water solubility, vapor pressure, and octanol-water partition coefficient, with a view to using them later in calculations of environmental fate.

The structures of many of these chemicals are given in Figure 3.1. Table 3.2 gives relevant properties of a group of "benchmark" chemicals selected to cover a wide range of properties. Several of these have been used by Neely and Blau (1985) in their two-volume text on environmental chemistry. These data have been selected from a number of sources including Verschueren (1983), Miller et al. (1985), Horvath and Getzen (1985), Shaw (1989), Mackay and Shiu (1981), Zwolinski and Wilhoit (1971), Hansch and Leo (1979), Callahan et al. (1979), Shiu and Mackay (1986), Shiu et al. (1988), Suntio et al. (1988), Yalkowsky (1989), and Friesen et al. (1990).

3.3.2   Chemical Classes

The Hydrocarbons

Hydrocarbons are naturally occurring chemicals present in crude oil and natural gas. Some are formed by biogenic processes in vegetation but most contamination comes from oil spills, and effluents from petroleum and petrochemical refineries, and from fuel use.

The alkanes can be separated into classes of normal and branched (or iso) species and cyclic alkanes, which range in molecular weight from methane or natural gas to waxes. They are usually sparingly soluble in water. For example, hexane has a solubility of approximately 10 g/m^3. This solubility falls by a factor of about 3 or 4 for every carbon added. The branched and cyclic alkanes tend to be more soluble in water, apparently because they have smaller molecular volumes.

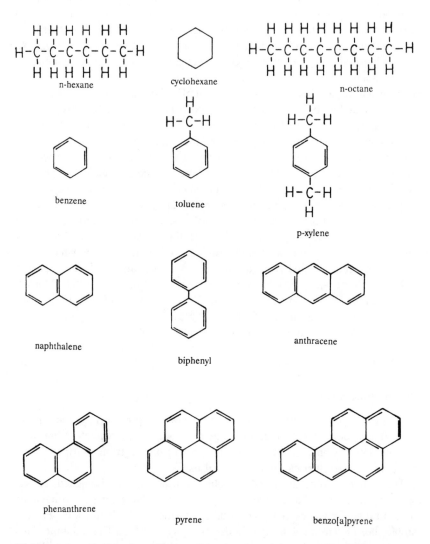

Figure 3.1.   Structures of selected chemicals of environmental interest.

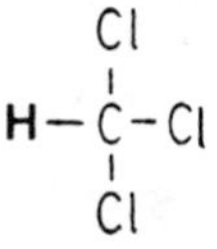

chloroform

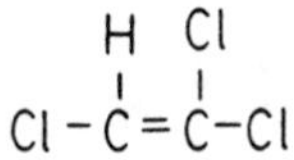

trichloroethylene

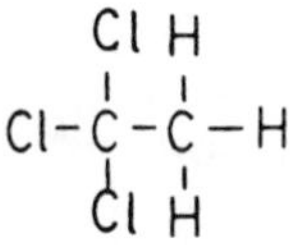

1,1,1-trichloroethane

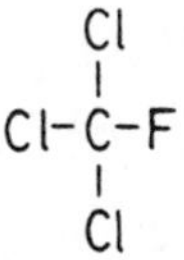

trichlorofluoromethane

chlorobenzene

1,4-dichlorobenzene

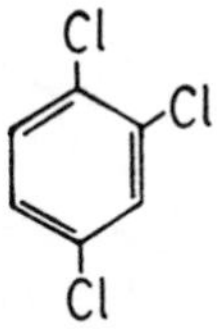

1,2,4-trichlorobenzene

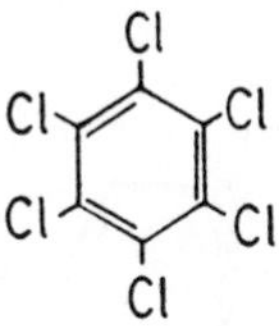

hexachlorobenzene

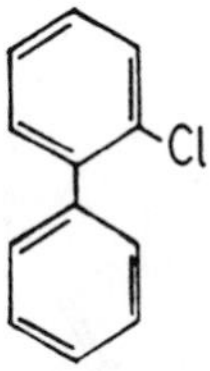

2-chlorobiphenyl

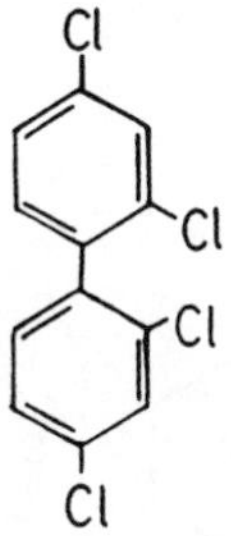

2,2'4,4'-tetrachlorobiphenyl

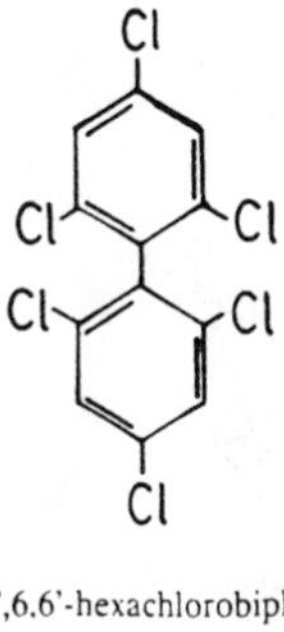

2,3,7,8-TCDD

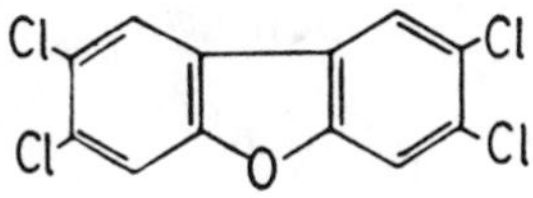

2,3,7,8-TCDF

2,2',4,4',6.6'-hexachlorobiphenyl

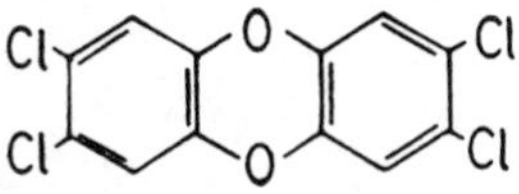

Figure 3.1.   Continued

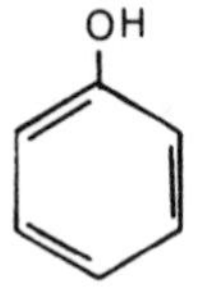

phenol

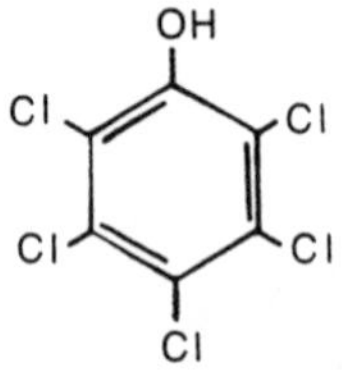

pentachlorophenol

p-cresol

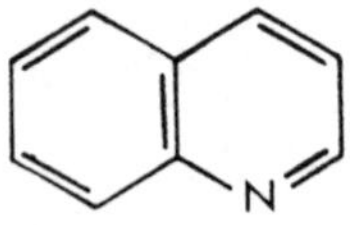

quinoline

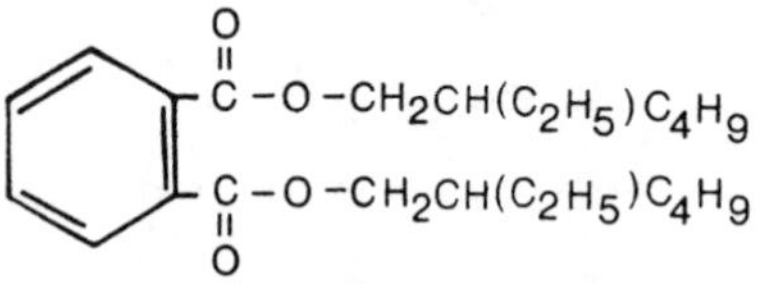

diethylhexylphthalate

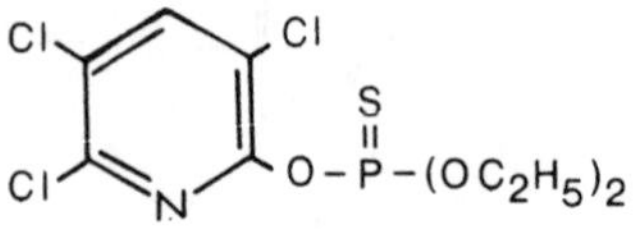

chlorpyrifos

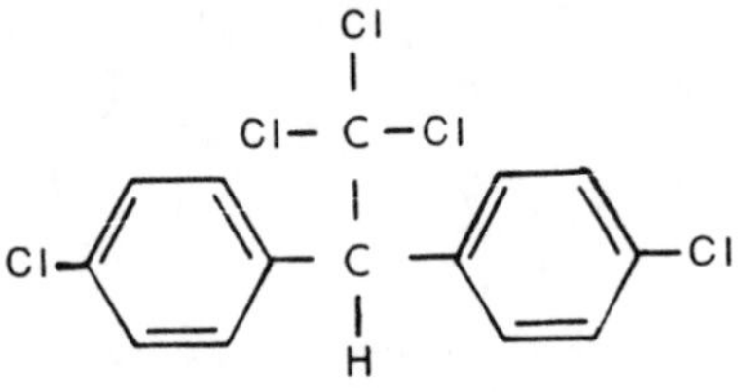

DDT

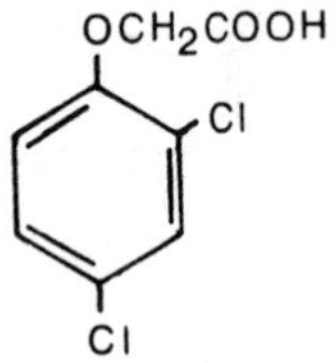

2,4-D

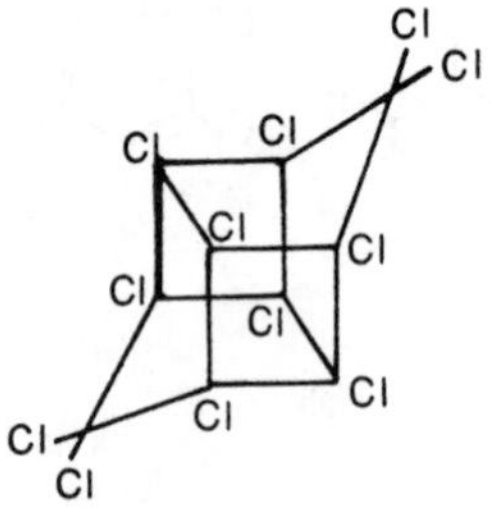

mirex

Figure 3.1.   Continued

Table 3.2. Physical-Chemical Properties of Selected Organic Chemicals at 25°C

Chemical	Molecular Mass, g/mol	mp, °C	bp, °C	Solubility g/m³	Vapor Press. Pa	log K_{ow}
n-Hexane	86.2	−95	68	9.5	20200	4.11
Cyclohexane	84.2	6.55	80.7	55	12700	3.44
n-Octane	114.2	−56.2	125.7	0.66	1880	5.18
Benzene	78.1	5.53	80	1780	12700	2.13
Toluene	92.1	−95	111	515	3800	2.69
p-Xylene	106.2	13.2	138	185	1170	3.15
Naphthalene	128.2	80.2	218	31.7	10.4	3.35
Biphenyl	154.2	71	277.5	7.48	1.2	4.03
Anthracene	178.2	216.2	340	0.041	0.0008	4.63
Phenanthrene	178.2	101	339	1.29	0.0161	4.57
Pyrene	202.3	156	360	0.135	0.0006	5.22
Benzo(a)pyrene	252.3	175		0.0038	0.0000007	6.04
Chloroform	119.4	−63.5	81	8200	23080	1.97
Trichloroethylene	131.4	−73	87	1100	9870	2.29
1,1,1-Trichloroethane	133.4	−32	113	730	12800	2.47
Trichlorofluoromethane	137.4	−111	23.8	1100	91600	2.53

(continued on next page)

Table 3.2.   Continued

Chemical	Molecular Mass, g/mol	mp, °C	bp, °C	Solubility g/m³	Vapor Press. Pa	log K_{ow}
Chlorobenzene	112.6	−46.5	132	472	1580	2.84
1,4-Dichlorobenzene	147.0	53.1	174	83.1	90.2	3.40
1,2,4-Trichlorobenzene	181.5	17	213.5	34.6	60.6	4.00
Hexachlorobenzene	284.8	230	322	0.005	0.0023	5.50
2-Chlorobiphenyl	188.7	34	374	1.3	2.04	4.54
2,2',4,4'-Tetrachlorobiphenyl	291.9	83		0.068	0.02	5.90
2,2',4,4',6,6'-Hexachlorobiphenyl	360.9	114		0.0007	0.0016	7.00
2,3,7,8-TCDD	322.0	305		0.0000193	0.0000001	6.80
2,3,7,8-TCDF	306.0	227		0.000419	0.000002	6.10
DDT	354.5	108.5		0.0031	0.00002	6.19
Mirex	545.6	485		0.00007	0.0001	6.89
Phenol	94.1	40.9	181.75	82000	70.6	1.46
Pentachlorophenol	266.4	190	310	14	0.0147	5.01
p-Cresol	108.1	34.8	201.9	16800	14.67	1.95
Quinoline	129.2	−15.6	237.7	60000	133	2.03
Diethylhexylphthalate	390.6	−50	386.9	0.4	0.0000267	5.30
Chlorpyrifos	350.6	42		0.4	0.0015	5.11
2,4-D (acid)	221.0	138	215	890	0.000056	2.81

Highly branched or cyclic alkanes such as terpenes are produced by vegetation. They are often sweet smelling and tend to be very resistant to biodegradation.

The alkenes or olefins are not naturally occurring to any significant extent. They are mainly used as petrochemical intermediates. The alkynes, of which acetylene is the first member, are also chemical intermediates which are rarely found in the environment. These unsaturated hydrocarbons tend to be fairly reactive and short-lived in the environment, whereas the alkanes are more stable and persistent.

Of particular environmental interest are the aromatics, the simplest of which is benzene. The aromatics are relatively soluble in water; for example, benzene has a solubility of 1780 g/m^3. They are regarded as fairly toxic and often troublesome compounds. A variety of substituted aromatics can be obtained by substituting various alkyl groups. For example, methyl benzene is toluene.

When two benzene rings are fused the result is naphthalene, which is also a chemical of considerable environmental interest. Subsequent fusing of benzene rings to naphthalene leads to a variety of chemicals referred to as the polycyclic aromatic hydrocarbons or polynuclear aromatic hydrocarbons, PAHs or PNAs. These compounds tend to be formed under conditions when a fuel is burned with insufficient oxygen. They are thus present in exhausts from engines, and are of interest because many are carcinogenic.

Biphenyl is a hydrocarbon which is not of much importance as such, but forms an interesting series of chlorinated compounds, the PCBs or polychlorinated biphenyls.

Halogenated Hydrocarbons

If the hydrogen in a hydrocarbon is substituted by chlorine, or less frequently by bromine, iodine, and fluorine, the resulting compound tends to be less flammable, more stable, more hydrophobic, and more troublesome environmentally. Replacing a hydrogen by a chlorine usually causes an increase in molecular volume and a corresponding decrease in solubility by a factor of about 3.

The stability of many of these compounds makes them useful as solvents, examples being chloroform, methylene chloride, and carbon tetrachloride. The fluorinated and chlorofluoro compounds are very stable and are used as refrigerants. Because these molecules are quite small, they are fairly soluble in water and are therefore able to penetrate the tissues of organisms quite readily. They are thus used as anaesthetics and narcotic agents.

The chlorinated aromatics are a particularly troublesome group of chemicals. The chlorobenzenes are biologically active. 1,4 or p-dichlorobenzene is widely used as a deodorant and disinfectant. The polychlorinated biphenyls or PCBs and their brominated cousins, the PBBs, are notorious environmental contaminants, as are chlorinated terpenes such as toxaphene, which is a very potent and long-lived insecticide. Many of the early pesticides such as DDT, mirex, chlordane, and lindane, are chlorinated hydrocarbons. They possess the desirable properties of stability, and a high tendency to partition out of air and water into the target organisms. Thus application of a pesticide results in protection for a prolonged time. As Rachel Carson demonstrated in *Silent Spring*, the problem is that these chemicals persist long enough to affect non-target organisms, and to drift throughout the environment causing widespread contamination.

Oxygenated Compounds

The most common oxygenated organic compounds are the alcohols, ethanol being among the most widely used. Others are octanol, which is a convenient analytical surrogate for fat, and glycerol, which is of interest because it forms the backbone of fat molecules by esterifying with fatty acids to form glycerides.

The phenols consist of an aromatic molecule in which a hydrogen is replaced by an OH group. They are acidic and tend to be biologically disruptive. Phenol, or carbolic acid, was the first disinfectant. Adding chlorines to phenol tends to increase the potency of the substance, and its tendency to ionize. Pentachloro-phenol (PCP) is a particularly toxic chemical, and is widely used for wood preservation.

The ketones, such as acetone, and aldehydes, such as formaldehyde, are fairly reactive in the environment and can be quite troublesome as atmospheric contaminants in regions close to sources of emission. Much of the problem of smog is attributable to aldehydes formed in combustion processes.

Organic acids such as acetic acid are also fairly reactive. They are not usually regarded as environmentally troublesome. Some chlorinated organic acids; for example, 2,4-D, are potent herbicides. Longer chain acids, such as stearic acid are mainly of interest because they esterify with glycerol to form fats. Humic and fulvic acids are of considerable environmental importance. These are substances of complex and variable structure which are naturally present in soils, water, and sediments. They can be regarded as the remnants of living organic materials such as wood which has been subjected to prolonged microbial conversion. These acids are sparingly soluble in water, but the solubility can be increased by the presence of alkali.

The esters or "salts" of organic acids and alcohols tend to be relatively in-nocuous and short-lived in most cases. They are subject to hydrolysis to the original components. A notable exception is the phthalate esters, which are very stable oily substances that are invaluable additives (plasticizers) for plastics, rendering them more flexible. Phthalate esters include the ester with two molecules of 2 ethylhexanol (DEHP). The other esters of interest are the glycerides; for example, glyceryl trioleate, the ester of glycerine and oleic acid. This chemical is very similar in properties to fat and has been suggested as a convenient surrogate for measuring fat to water partitioning.

The "dioxins" are a series of organic compounds which have recently become environmentally notorious. The chlorinated dibenzo-p-dioxins are formed under combustion conditions when chlorine is present. They form a series of very toxic chemicals, the most celebrated of which is 2,3,7,8-tetrachlorodibenzo-p-dioxin (TCDD). TCDD is possibly the most toxic chemical to mammals. A dose of 2 μg of TCDD per kg of body weight is sufficient to kill small rodents.

A related series of chemicals is the dibenzofurans, which are similar in properties to the dioxins. It appears that molecules which are long and flat and have chlorine atoms strategically located at the ends are particularly troublesome environmentally. Examples are the chloronaphthalenes, DDT, the PCBs, and chlorinated dibenzo-p-dioxins and dibenzofurans.

Other oxygenated compounds of interest include the carbohydrates, cellulose, and lignins which occur naturally.

Nitrogen Compounds

Nitrogen compounds of environmental interest include the amines, amides, pyridines, quinolines, the amino acids, various nitro compounds including nitro PNAs and nitroso compounds. Many of these compounds occur naturally, are quite toxic, and are difficult to analyze.

Sulfur Compounds

Sulfur compounds including thiols, thiophenes, and mercaptans are well-known because of their strong odor. One of the most prevalent classes of synthetic organic chemicals is the alkyl benzene sulfonates which is widely used in detergents.

Phosphorus Compounds

Phosphorus compounds play a key role in energy transfer in organisms. Organophosphate compounds have been developed as pesticides; for example, chlorpyrifos, which have the very desirable properties of high biological activity but relatively short environmental persistence. They have therefore largely replaced organochlorine compounds in agriculture.

Arsenic Compounds

Arsenic, which behaves somewhat similarly to phosphorus, is inadvertently produced in mineral processing, and has a long and celebrated history as a poison. It usually exists in anionic form.

Metals

Most metals are essential for human life in small quantities but can be toxic if administered in excessive dosages. The metals of primary toxicological interest here are those which form organometallic molecules. Notable is mercury which can exist as the element in various ionic and organometallic forms. Other metals such as lead and tin behave similarly. A formidable literature exists on the behavior, fate, and effects of the "heavy" metals such as lead, copper, and chromium. These metals often have a complex environmental chemistry and toxicology which varies considerably, depending on their ionic state as influenced by acidity and redox status.

Other Chemicals and Pollution Parameters

Several other chemicals are of environmental concern, including ozone, radon, chlorine, organic and inorganic sulfides, and cyanides, as well as the indeterminate broad class of "conventional" pollutants such as biochemical oxygen demand (BOD) and chemical oxygen demand (COD). Finally, certain mineral substances such as asbestos are of concern, more because of their physical structure than their chemical composition.

The Future

It would be unwise to assume that current lists of priority chemicals are complete and will remain static. To a large extent the chemicals on the lists reflect our ability to detect and analyze them, rather than their real significance. The prevalence of organochlorine chemicals on lists is largely the result of availability of the sensitive electron capture detector. As new analytical methods emerge, new chemicals will be found and priorities will change. Happy hunting grounds for environmental chemists include combustion gases, dyes, mine tailings, effluents from pulp and paper operations (especially those involving chlorine bleaching), landfill leachates, and a vast assortment of products of metabolic conversion in organisms ranging from bacteria to humans.

4   THE NATURE OF
ENVIRONMENTAL MEDIA

4.1 INTRODUCTION

The objective of this chapter is to present a qualitative description of environmental media, highlighting some of their more important properties. In doing so, it is useful to assemble "evaluative" environments which can be used later in calculations. We can consider, for example, an area 1 km by 1 km, consisting of some air, water, soil, and sediment. Volumes and properties can be assigned to these media which are typical, but purely illustrative, and will of course require modification if chemical fate in a specific region is to be treated. We examine in sequence, the atmosphere, the hydrosphere (i.e., water), then the lithosphere (bottom sediments and terrestrial soils), each with its resident biotic community.

It transpires that it is convenient to define two evaluative environments. First is a simple four compartment system which is easily understood and illustrates the general principles which are applied in multimedia calculations. Second is a more complex eight compartment system which is more representative of real environments, and is correspondingly more demanding of data, and leads to more lengthy calculations.

These environments are depicted in Figure 4.1. For a fuller account of these environments or "unit worlds" the reader can consult the discussion by Neely and Mackay (1982).

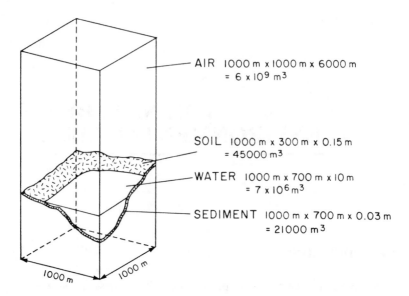

AIR 1000 m x 1000 m x 6000 m
     = 6 x 10^9 m^3

SOIL 1000 m x 300 m x 0.15 m
      = 45000 m^3

WATER 1000 m x 700 m x 10 m
       = 7 x 10^6 m^3

SEDIMENT 1000 m x 700 m x 0.03 m
          = 21000 m^3

1000 m   1000 m

SIMPLE FOUR COMPARTMENT ENVIRONMENT

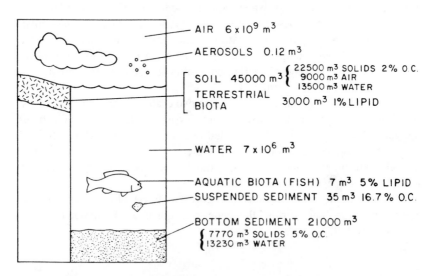

AIR   6 x 10^9 m^3

AEROSOLS   0.12 m^3

SOIL 45000 m^3 { 22500 m^3 SOLIDS 2% O.C.
                     9000 m^3 AIR
                     13500 m^3 WATER

TERRESTRIAL
BIOTA   3000 m^3 1% LIPID

WATER   7 x 10^6 m^3

AQUATIC BIOTA (FISH) 7 m^3 5% LIPID

SUSPENDED SEDIMENT 35 m^3 16.7% O.C.

BOTTOM SEDIMENT 21000 m^3
          { 7770 m^3 SOLIDS 5% O.C.
            13230 m^3 WATER

EIGHT COMPARTMENT ENVIRONMENT

Figure 4.1.   Evaluative environments.

4.2   THE ATMOSPHERE

4.2.1   Air

The layer of the atmosphere which is in most intimate contact with the surface of the earth is the troposphere, which extends to a height of about 10
km. The temperature, density, and pressure of the atmosphere fall steadily
with increasing height, which is a nuisance in subsequent calculations. If we
assume uniform density at atmospheric pressure, then the entire troposphere
can be viewed as being compressed into a height of about 6 km. Exchange
of matter from the troposphere through the tropopause to the stratosphere is
a relatively slow process and is rarely important in environmental calculations,
except in the case of chemicals such as the freons, which catalyze the destruction of stratosphere ozone, thus facilitating the penetration of UV light to the
Earth's surface. A reasonable atmospheric volume over our 1 km square world
is thus $1000 \times 1000 \times 6000$ or 6×10^9 m^3.

If our environmental model is concerned with a localized situation, for example, a state, province, or metropolitan region,it is unlikely that most pollutants would manage to penetrate higher than about 1000 to 2000 m during the
time the air resides over the region. It may therefore be appropriate to reduce
the height of the atmosphere to 1000 to 2000 m in such cases. In extreme cases,
for example over small ponds or fields, the accessible mixed height of the atmosphere may be as low as 10 m. The modeler must make a judgment as to
the volume of air which is accessible to the chemical during the time that the
air resides in the region of interest.

4.2.2   Aerosols

The atmosphere contains a considerable amount of particulate matter or aerosols which are important in determining the fate of certain chemicals. These
particles may range from water in the form of fog or cloud droplets to dust
particles from soil and smoke from combustion. The concentration of these
aerosols is normally reported in $\mu g/m^3$. A rural area may have a concentration of about 5 $\mu g/m^3$, and a fairly polluted urban area a concentration of
100 $\mu g/m^3$. For illustrative purposes we can assume that the particles have
a density of 1.5 g/cm^3 and are present at a concentration of 30 $\mu g/m^3$. This
corresponds to volume fraction of particles of 2×10^{-11}. The density of these
particles is usually unknown; thus the volume fractions are only estimates.
It is, however, convenient for us to calculate this amount in volume fraction
form. In an evaluative air volume of 6×10^9 cubic meters, there is thus 0.12
cubic meters or 120 liters of solid material.

4.2.3 Deposition Processes

This aerosol material has a very high surface area and thus absorbs, or adsorbs, or sorbs, many pollutants, especially those of very low vapor pressure such as the PCBs or polyaromatic hydrocarbons. In the case of benzo(a)pyrene, almost all the chemical present is associated with particles, and very little exists in the gas phase. This is important because chemical associated with aerosol particles is subject to two important deposition processes. First is dry deposition, in which the aerosol particle can be regarded (somewhat erroneously) as falling under the influence of gravity to the earth's surface. This falling velocity, or deposition velocity is quite slow and depends on the condition of atmosphere, the size and properties of the aerosol particle, and the nature of the ground surface, but a typical velocity is about 0.3 cm/s or 10.8 m/h. The result is deposition of 10.8 m/h $\times$ 2 $\times$ 10^{-11} (volume fraction) $\times$ 10^6 m^2 or 0.000216 m^3/h or 1.89 m^3/year. Second, the particles may be scavenged or swept out of the air by wet deposition with rain drops. As it falls, each rain drop sweeps through a volume of air about 200,000 times its volume prior to landing on the land or water surface. It thus has the potential to remove a considerable quantity of aerosol from the atmosphere. Rain is, therefore, often highly contaminated with substances such as PCBs and PAHs. There is a common fallacy that rain water is pure. In reality it is usually much more contaminated than surface water. Typical rainfall rates lie in the range 0.3 to 1 meter per year, but of course, vary greatly with climate. We adopt a figure of 0.8 m/year for illustrative purposes. This results in the scavenging of 200,000 $\times$ 0.8 m/year $\times$ 2 $\times$ 10^{-11} $\times$ 10^6 m^2 or 3.2 m^3/year of aerosol particles, about twice the rate of dry deposition.

In the four compartment evaluative environment we ignore aerosols, but we include them in the eight compartment version.

4.3 THE HYDROSPHERE OR WATER

4.3.1 Water

Some 70% of the Earth's surface is covered by water. In some evaluative models, the area of water is taken as 70% of the million square meters or 700,000 m^2. Similarly to the atmosphere, only near-surface water is accessible to pollutants in the short term. In the oceans, this depth is about 100 meters, but since most situations of environmental interest involve fresh or estuarine water, it is more appropriate to use a shallower water depth of perhaps 10 meters. This yields a water volume of about 7 $\times$ 10^6 m^3. If the aim is to mimic the proportions of water and soil in a political jurisdiction, such as a state or province, the area of water will normally be considerably reduced to perhaps 15% of the total, giving a volume of about 10^6 m^3. We normally regard the water as being pure, i.e., containing no dissolved electrolytes, but we do treat its content of suspended particles.

4.3.2 Particulate Matter

Particulate matter in the water plays a key role in influencing the behavior of chemicals. Again, we do not normally know if the chemical is absorbed or adsorbed to the particles. We play safe and use the vague term "sorbed." A very clear water may have a concentration of particles as low as 1 g/m^3, or the equivalent 1 mg/L. However, in most cases the concentration is higher, in the range of 5 to 20 g/m^3. Very turbid, muddy waters may contain over 100 g/m^3. Assuming a concentration of 7.5 g/m^3 and a density of 1.5 g/cm^3 gives a volume fraction of particles of about 5×10^{-6}. Thus in the 7×10^6 m^3 of water, there are 35 m^3 of particles.

This particulate matter consists of a wide variety of materials. It contains mineral matter which may be clay or silica in nature. It also contains dead or detrital organic matter which is often referred to as humin, humic acids, and fulvic acids or, more vaguely, as organic matter. It is relatively easy to measure the total concentration of organic carbon (OC) in water or particles by converting the carbon to carbon dioxide and measuring the amount spectroscopically. Alternatively, the solids can be dried to remove water, then heated to ignition temperatures to burn off organic matter. The loss is referred to as "loss on ignition" (LOI) or as "organic matter" (OM). Thus, there are frequent reports of the amount of dissolved organic carbon (DOC) or total organic carbon (TOC) in water. These humic and fulvic acids have been the subject of intense study for many years. They are organic materials of variable composition which probably originate from the ligneous material present in vegetation. They contain a variety of chemical structures including substituted alkane, cycloalkane, and aromatic groups, and have acidic properties imparted by phenolic and carboxylic acids. They are therefore fairly soluble in alkaline solution, in which they are present in phenolate or carboxylate ionic form, but they may be precipitated under acidic conditions. The operational difference between humic and fulvic acids is the pH at which precipitation occurs.

It is important to avoid confusing organic matter (OM) and organic carbon (OC). Typically OM contains 50% to 60% OC, thus an OM analysis of 10% may also be 5% OC. A mass basis, i.e., g/100 g dry particles is commonly used. For convenience in our evaluative calculations we will usually treat OM as 50% OC, and we will assume a density for both OM and OC as being equal to that of water.

Concentrations of these suspended materials may be defined operationally using filters of various pore size; for example, 0.45 μm. There is a tendency to describe material which is smaller than this, i.e., which passes through the filter, as being operationally "dissolved." It is not clear how we can best discriminate between "dissolved" and "particulate" forms of such material since there is presumably a continuous size spectrum ranging from molecules of a few nm, to relatively large particles of 100 or 1000 nm. It transpires that the organic material in the suspended phases is of great importance because it has a high sorptive capacity for organic chemicals. It is therefore common to assign an organic carbon content to these particulate phases. In a fairly productive lake the OM content of particles may be as high as 50%, but for illustrative purposes a figure of 33% for OM or 16.6% OC is convenient. In each cubic meter of water there is thus 2.5 g or cm^3 of OM, and 5.0 g or 2.5 cm^3 of mineral matter of density 2 g/cm^3, totaling 7.5 g or 5.0 cm^3, giving an average particle density of 1.5 g/cm^3.

Also present in the water is a variety of organisms or aquatic biota ranging from bacteria to algae, invertebrates, and fish.

4.3.3 Fish and Aquatic Biota

Fish are of particular interest because they are of commercial and recreational importance to users of water, and they tend to bioconcentrate or bioaccumulate metals and organic chemicals from water. They are thus convenient monitors of the contamination status of lakes. This raises the question, "What is the volume fraction of fish in a lake?" Most anglers and even aquatic biologists greatly overestimate this number. It is probably, in most cases, in the region of 10^{-8}, but this is somewhat misleading because most of the biotic material in a lake is not fish, it is material of lower trophic levels, on which fish feed. For illustrative purposes, we can assume that all the material in the water is fish and the total concentration is about 1 part per million by volume, yielding a volume of "fish" of about 7 m^3. Later, it proves useful to define a lipid or fat content of fish, a figure of 5% by volume being typical.

In summary, the water thus consists of 7×10^6 m^3 of water, containing 35 m^3 of particulate matter and 7 m^3 of "fish" or biota.

In shallow or near-shore water there may be a considerable quantity of aquatic plants or macrophytes. These plants provide a substrate for a thriving microbial community and they possess inherent sorptive capacity. Their importance is usually underestimated. Because of the present limited ability to quantify their sorptive properties, we ignore them.

4.3.4 Deposition Processes

The particulate matter in water is important because, like aerosols in the atmosphere, it serves as a vehicle for the transport of chemical from the bulk of the water to the bottom sediments. Hydrophobic substances tend to partition appreciably onto water particles, and are thus subject to fairly rapid deposition. This deposition velocity is typically 0.5 to 2.0 meter per day or 0.02 to 0.08 m/h. This velocity is sufficient to cause removal of most of the suspended matter from most lakes during the course of a year. Thus, under ice covered lakes in the winter, the water may clarify. Some of the deposited particulate matter is resuspended from the bottom sediment by the action of currents, storms, and the disturbances caused by bottom dwelling fish and invertebrates. During the summer there is considerable photosynthetic fixation of carbon by algae, resulting in the formation of large quantities of organic carbon in the water column. Much of this is destined to fall to the bottom sediments, but much is degraded by microorganisms within the water column.

Assuming, as discussed earlier, a figure of 5×10^{-6} m^3 particles per m^3 of water and a deposition velocity of 200 m per year, the deposition rate will be 0.001 m^3 of particles per m^2 of sediment area per year, or for an area of 7×10^5 m^2, a flow of 700 m^3/year. We examine this rate in more detail in the next section.

4.4 BOTTOM SEDIMENTS

4.4.1 Sediment Solids

Inspection of the state of the bottom of lakes reveals that there is a fairly fluffy or nepheloid active layer at the water/sediment interface. This layer typically consists of 95% water and 5% particles, and is often highly organic in nature. It may consist of deposited particles and fecal material from the water column. It is stirred by currents, and by the action of the various biota present in this "benthic" region. The sediment becomes more consolidated at greater depths, and the water content tends to drop toward 50%. The top few centimeters of sediment are occupied by burrowing organisms which feed on the organic matter and on each other, and generally turn over (bioturbate) this entire "active layer" of sediment. Depending on the condition of the water column above, this layer may be oxygenated (aerobic or oxic) or depleted of oxygen (anaerobic or anoxic). This has profound implications for the fate of inorganic substances such as metals and arsenic but it is relatively unimportant for organic chemicals, except in that the oxygen status influences the nature of the microbial community, which in turn influences the availability of metabolic pathways for chemical degradation. The deeper sediments are less accessible, and ultimately the material (and its chemical) becomes almost completely buried and inaccessible to the aquatic environment above. Most of the activity occurs in the top 5 cm of the sediment but it is misleading to assume that sediments deeper than this are not accessible. There remains a possibility of bioturbation or diffusion.

Bottom sediments are difficult to investigate; they are unpleasant and have little or no commercial value. They are thus often ignored. This is unfortunate because they serve as the depositories for much of the toxic material discharged into water. They are thus very important and valuable and merit more sympathy and attention.

Rivers are normally so turbulent that the bottom is scoured, exposing rock or consolidated mineral matter; thus the sediments are less important.

4.4.2   Deposition, Resuspension and Burial

It is possible to estimate the rate of deposition, i.e., the amount of material which is falling annually to the bottom of the lake and is being retained there. This can be done by sediment traps which are essentially trays which collect falling particles, or by taking a sediment core and assigning dates to material at various depths using concentrations of various radioactive metals, such as lead. Nuclear events also provide convenient dating markers for sediment depths. The measurement of deposition is complicated by the presence of the simultaneous reverse process of resuspension caused by currents and biotic activity. It is difficult to measure how much material is rising and falling since much may be merely cycling up and down in the water column. Burial or net deposition rates vary enormously but a figure of about 1 mm per year is typical. Much of this is water, which is trapped in the burial process.

Chemicals present in sediments are primarily removed by degradation, burial, or resuspension and diffusion back to the water column.

For illustrative purposes we adopt a sediment depth of 3 cm and suggest that it consists of 63% water, 37% solids by volume and these solids consist of about 10% organic matter or 5% organic carbon. Living creatures are included in this figure.

Some of this deposited organic material is resuspended to the water column, some of it is degraded, i.e., it is used as a source of energy by benthic or bottom-living organisms, and some is destined to be permanently buried. The low 5% organic carbon figure for deeper sediments compared to the high 17%

for the depositing material implies that about 75% of the organic carbon is degraded.

It is now possible to assemble an approximate mass balance for the sediment mineral matter (MM), organic matter (OM), and thus the organic carbon (OC). Table 4.1 illustrates this mass balance.

On a 1 square meter basis, the deposition rate may be 0.001 m^3 per year or 1000 cm^3 per year, which with a particle density of 1.5 g/cm^3 corresponds to 1500 g/year, of which 500 g is OM and 1000 g is MM. We assume that 40% of this is resuspended, i.e., 200 g of OM and 400 g of MM. Of the remaining 300 g OM, we assume that 233 g is degraded to CO_2 and 67 g is buried along with the remaining 600 g of MM. Total burial is thus 667 g, which consists of 600 g or 300 cm^3 of MM and 67 cm^3 of OM, i.e., 10% OM by mass or 18% OM by volume. The total volumetric burial rate of solids is 367 cm^3/year. Now associated with these solids is 633 cm^3 of pore water, thus the total volumetric burial rate of solids plus water is approximately 1000 cm^3/year, corresponding to a rise in the sediment-water interface of 1 mm/year. The percentage of OC in the depositing and resuspending material is 17% while in the buried material it is 5%. The bottom sediment bulk density, including pore water, is thus 1300 kg/m^3.

On a 7×10^5 m^2 basis the deposition rate is 700 m^3/year, resuspension is 280 m^3/year, burial is 257 m^3/year, and degradation accounts for the remaining 163 m^3/year.

The organic and mineral matter balances are thus fairly complicated, but it is important to define them because they control the fate of many hydrophobic chemicals.

It is noteworthy that the burial rate of 1 mm/year coupled to the sediment depth of 3 cm indicates that on the average it will take 30 years for sediment solids to become buried. During this time they may continue to release sorbed chemical back to the water column. This is the crux of the "in place contaminated sediments" problem which is unfortunately very common, especially in the Great Lakes Basin.

In the simple four-compartment environment we treat only the solids, but in the eight-compartment version we include the sediment pore water. In the interests of simplicity we assign a density of 1500 kg/m^3 to the sediment in the four compartment model.

4.4.3   Diffusion from Sediments

This already complicated picture of transport and transformation becomes even more complex when attempts are made to calculate rates of diffusion

Table 4.1. Sediment-Water Mass Balance on a 1 m² Area Basis

	Mineral Matter cm³	g	Organic Matter cm³	g	Total cm³	g	Organic Carbon g
Deposition	500	1000	500	500	1000	1500	250
Resuspension	200	400	200	200	400	600	100
OM conversion	—	—	233	233	233	233	167
Burial (solids)	300	600	67	67	367	667	33
Buried water	—	—	—	—	633	633	—

Total burial is 1000 cm³/year or 1300 g/year corresponding to a "velocity" of 1 mm/year.

The sediment thus has a density of 1.300 g/cm³ or 1300 kg/m³

Assumed densities are: Mineral Matter 2 g/cm³
                         Organic Matter 1 g/cm³

Organic matter is 50% (mass) organic carbon.

of chemicals from sediments. The pore water or interstitial water in sediments contains colloidal organic matter to which chemical is sorbed. Chemical dissolved in the pore water, or sorbed to these colloids, can migrate or diffuse from the sediments upward to the water column or downward to greater depths. Often the amount on the colloids greatly exceeds the amount dissolved. Finally, there may be vertical upward or downward flow of water through the sediments induced by hydrostatic differences. We discuss these phenomena later in Chapter 8.

4.5 SOILS

4.5.1 The Nature of Soil

Soil is a complex organic matrix consisting of air, water, mineral matter, notably clay and silica, and organic matter, which is similar in general nature to the organic matter discussed earlier for the water column.

The surface soil is subject to diurnal and seasonal temperature changes and to marked variations in water content, and thus in air content. At times it may be completely flooded, and at other times almost completely dry. The organic matter in the soil plays a crucial role in controlling the retention of the water, and thus in ensuring the viability of plants. The organic matter content is typically 1% to 5%. Depletion of organic matter through excessive agriculture tends to render the soil infertile, an issue of great concern in agricultural regions. Soils vary enormously in their composition and texture and consist of various layers, or horizons, of different properties. There is transport vertically and horizontally by diffusion in air and in water, flow or advection in water, and of course movement of water and nutrients into plant roots and thence into stems and foliage. Burrowing animals can also play an important role in transporting chemicals in soils.

A typical soil may consist of 50% solid matter, 20% air, and 30% water, by volume. The solid matter may consist of about 2% organic carbon or 4% organic matter. During and after rainfall, water flows vertically downward through the soil and may carry chemicals with it. During periods of dry weather, water often returns to the surface by capillary action, again moving the chemicals.

Most soils are, of course, covered with vegetation, which stabilizes the soil and prevents it being eroded by wind or water action. Under dry conditions, with poor vegetation cover, considerable quantities of soil can be eroded by wind action, carrying with it adsorbed chemicals. Sand dunes are an extreme example. In populated regions the loss of soil in water runoff is of more concern. This water often contains very high concentrations of soil, perhaps as much as a volume fraction of 1 part per thousand of solid material. This serves as a vehicle for the movement of chemicals, especially agricultural chemicals such as pesticides, from the soils to water bodies such as lakes.

In most areas there is a net movement of water vertically from the surface soil to greater depths into a pervious layer of rock or aquifer through which groundwater flows. The quality of this groundwater has become of considerable concern recently, especially to those who rely on wells for water supply. This water tends to move very slowly, i.e., at a velocity of meters per year, through the porous subsurface strata. If contaminated, it can take decades or even centuries to recover. Of particular concern are regions in which chemicals have seeped from dumps or landfills into the groundwater and have migrated some distance into rivers, wells, or lakes. It is quite difficult and expensive to investigate, sample, and measure contaminant flow in groundwater. It may even not be obvious in which direction the water is flowing, and how fast it is flowing. Chemicals associated with groundwater generally move more slowly than the velocity of the groundwater. They are retarded by a "retardation factor" which is essentially the ratio of the amount of chemical which is sorbed to the solid matrix, to the amount which is in solution. Sorption of organic chemicals is usually preferentially to organic matter; however, clays also have considerable sorptive capacity, especially when dry. The characterization of migration of chemicals in groundwater is difficult, and especially so when a chemical is present in a nonaqueous phase; for example, as a bulk oil or emulsified oil phase.

For illustrative purposes, we treat the soil as covering an area 1000 m by 300 m by 15 cm deep, which is about the depth to which agricultural soils are plowed. This yields a volume of 45,000 m³. The soil consists of about 50% solids of 4% organic matter content, or 2% by mass organic carbon. The porosity of the soil, or void space, is 50%, and consists of 20% air and 30% water. Assuming a density of the soil solids of 2400 kg/m³ and water of 1000 kg/m³ gives masses of 1200 kg solids and 300 kg water per m³ (and a negligible 0.2 kg air), totaling 1500 kg, corresponding to a bulk density of 1500 kg/m³. Rainwater falls on this soil at a rate of 0.8 m per year. Of this, perhaps 0.3 m evaporates, 0.3 m runs off, and 0.2 m percolates to depths and contributes to groundwater flow. This results in water flows of 90,000 m³/year by evaporation, 90,000 m³/year by runoff and 60,000 m³/year by percolation to depths, totaling 240,000 m³/year. With the runoff is associated 90 m³/year of solids, i.e., a concentration of 0.1% by volume. Again it must be emphasized that these numbers are entirely illustrative.

This soil runoff rate of 90 m³/year does not correspond to the sediment deposition rate of 700 m³/year, partly because of the contribution of organic matter generated in the water column, but mainly because of the low ratio of soil area to water area.

4.5.2   Terrestrial Biota or Plants

Until recently, most environmental models have ignored terrestrial vegetation. The reason for this is not that vegetation is unimportant, but rather that until recently modelers have had difficulty calculating the partitioning of chemicals into plants. This topic is now receiving more attention as a result of the realization that consumption of contaminated vegetation, whether by humans, domestic animals, or wildlife, is a major route or vector for transfer of toxic chemicals from one species to another, and ultimately to humans. Plants play a critical role in stabilizing soils, in inducing water movement from soil to the atmosphere, and they may serve as collectors and recipients of toxic chemicals deposited or absorbed from the atmosphere. Much of the credit for recognizing the importance of vegetation is due to the studies of Calamari, Bacci, and Vighi (1987) in Italy.

Amounts of biomass per m^2 vary enormously from near zero in deserts to massive quantities which greatly exceed soil volumes in tropical rain forests. They also vary seasonally. If it is desired to include vegetation, a typical "depth" of plant biomass might be 1 cm. This, of course, consists mainly of water, cellulose, starch, and ligneous material. There is little doubt that future, more sophisticated models will include more detailed descriptions of chemical partitioning behavior into plants. At present it is convenient (and admittedly unsatisfactory) to regard the plants as having a volume of 3000 m^3, containing the equivalent of 1% lipid-like material and 50% water.

We ignore plants in the simple four compartment model, treating the soil as only a simple solid phase.

4.6   SUMMARY

These evaluative volumes, areas, compositions and flowrates are summarized in Table 4.2. From them is derived a simple four compartment version. Also suggested is an alternative environment which is more terrestrial, less aquatic, and reflects more faithfully a typical political jurisdiction. It is emphasized again that the quantities are purely illustrative and site-specific values may be quite different. All that is needed at this stage is a reasonable basis for calculation.

Scientists who have devoted their lives to studying the intricacies of the structure, composition, and processes of the atmosphere or hydrosphere or lithosphere will undoubtedly be offended at the simplistic approach taken in

Table 4.2. Eight Compartment Evaluative Environment with an Alternative Environment with Less Water Given in Parentheses

Compartment	Volume m³	Density kg/m³	Composition
Air	6×10^9 (1×10^9)	1.2	–
Water	7×10^6 (1×10^6)	1000	–
Soil (50% solids, 20% air, 30% water)	4.5×10^4 (13.5×10^4)	1500	2% OC
Sediment (37% solids)	2.1×10^4 (3000)	1300	5% OC
Suspended Sediment	35 (5)	1500	16.7% OC
Aerosols	.12 (0.02)	1500	30 $\mu g/m^3$ or 2×10^{-11} vol. frn.
Aquatic Biota	7 (1)	1000	5% lipid
Terrestrial Biota	3000 (9000)	1000	1% lipid

Areas m²

Air-water　　　7×10^5 (1×10^5)
Water-sediment　7×10^5 (1×10^5)
Soil-air　　　　3×10^5 (9×10^5)

Process rates

Rain Rate　　0.8 m/year or 800,000 m³/year
560,000 m³ (80,000 m³) to water
240,000 m³ (720,000 m³) to soil

Aerosol Deposition Rates (total)

Dry deposition　216×10^{-6} m³/h or 1.89 m³/year
Wet deposition　365×10^{-6} m³/h or 3.2 m³/year

(continued on next page)

Table 4.2.   (Continued)

Sediment Deposition Rates

Deposition	700 m³/year (100 m³/year) solids	17% OC
Resuspension	280 m³/year (40 m³/year) solids	17% OC
Net deposition or burial	257 m³/year (38 m³/year) solids	5% OC

Fate of Water in Soil

Evaporation	90,000 m³/year	(270,000 m³/year)
Run off to water	90,000 m³/year	(270,000 m³/year)
Percolation to groundwater	60,000 m³/year	(180,000 m³/year)
Solids run off	90 m³/year	(270 m³/year)

Simple Four Compartment Environment

	Volume m³	Composition	Density kg/m³
Air	6×10^9	–	1.2
Water	7×10^6	–	1000
Soil	4.5×10^4	2% OC	1500
Sediment	2.1×10^4	5% OC	1500

this chapter. The environment is very complex, and it is essential to probe the fine detail present in its many compartments. But if we are to attempt broad calculations of multimedia chemical fate, we must suppress much of the media-specific detail. When the broad patterns of chemical behavior are established, it may be appropriate to revisit the media which are important for that chemical and focus on detailed behavior in a specific medium. At that time a more detailed and site-specific description of the medium of interest will be justified and needed.

5   PHASE EQUILIBRIUM

5.1   INTRODUCTION

We introduce the concept of equilibrium distribution of a chemical between phases by considering a simple two compartment system as illustrated in Figure 5.1.

A small volume of nonaqueous phase such as a particle of organic or mineral matter, a fish, or an air bubble is introduced into water which contains dissolved chemical such as benzene. There is a tendency for some of the benzene to migrate into this new phase and establish a concentration which is some multiple of that in the water. In the case of organic particles the multiple may be 100, or if the phase is air the multiple may be only 0.2. Equilibrium becomes established in hours or days between the benzene dissolved in the water and that in, or on, the nonaqueous phase. Analytical measurements may give the total or average concentration which includes the nonaqueous phase and may differ considerably from the actual water concentration. The phase may subsequently settle to the lake bottom, or rise to the surface, conveying benzene with it. Clearly, it is essential to establish the capability of calculating these concentrations, and therefore the fractions of the total amount of benzene which remain in the water, and enter the second phase. In some cases 95% of the benzene may migrate into the phase and in others only 5%. These systems will behave quite differently.

This chapter is devoted to developing methods of calculating such phase equilibria. It addresses the question, "Given a concentration in one phase, what will be the concentration in another phase which has been in contact with it long enough to achieve equilibrium?" This task is part of the science of thermodynamics which is fully described in several excellent texts, such as those of Denbigh (1966), Van Ness and Abbott (1982), Prausnitz et al. (1986), and for aquatic environmental systems by Stumm and Morgan (1981). It is

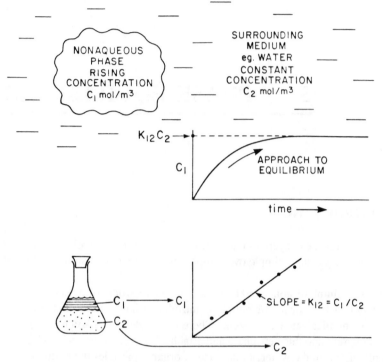

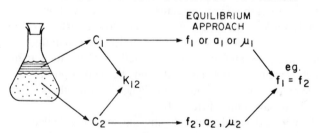

Figure 5.1. Some principles and concepts in phase equilibrium.

assumed here that the reader is familiar with the general principles of thermodynamics; thus, no attempt is made to derive the equations completely and

rigorously. The aim is rather to extract from the science of thermodynamics those parts which are pertinent to environmental chemical equilibria and explain their source, significance and applications.

Despite its name, thermodynamics is not concerned with process rates, indeed none of the equations derived in this chapter need contain time as a dimension. Rates are discussed in later chapters.

Environmental thermodynamics applies to a relatively narrow range of conditions. Temperatures range only between -40 and $+40°C$ and usually between the narrower limits of $0°$ and $25°C$. Total pressures are almost invariably atmospheric, but with, of course, an additional hydrostatic pressure at lake or ocean bottoms. Concentrations of chemical contaminants are (fortunately) usually low. Situations in which the concentration is high (as in spills of oil or chemicals) are best treated separately. These limited ranges are fortunate in that they simplify the equations and permit us to ignore large and complex areas of thermodynamics which treat behavior at high and low pressures and temperatures, and high concentrations.

The presence of a chemical in the environment rarely affects the overall dominant structure, processes, and properties of the environment; thus, we can take the environment "as is" and explore the behavior of chemicals in it with little fear of the environment being changed in the short term as a result. There are, however, certain notable exceptions, particularly when the biosphere (which can be significantly altered by chemicals) plays an important role in determining landform. An example is the stabilizing influence of vegetation on soils. Another exception is the role of deposited carbon of photosynthetic origin in lakes.

A point worth emphasizing is that thermodynamics is based on a few fundamental "laws" or axioms, from which an assembly of equations can be derived which relate certain useful properties to each other. Examples are a relationship between vapor pressure and enthalpy of vaporization, or concentration and partial pressure. In some cases the role of thermodynamics is to suggest suitable empirical relationships. Thermodynamics never defines the actual value of a property such as the boiling point of benzene; such data must be obtained experimentally. We thus process experimental data using thermodynamic relationships. We must generate it by experimental measurement.

It transpires that there are two approaches which can be used to develop equations relating equilibrium concentrations to each other as shown in Figure 5.1. The simpler and most widely used is Nernst's Distribution Law which postulates that the concentration ratio C_1/C_2 is relatively constant and equal to a partition or distribution coefficient K_{12}. Thus C_1 can be calculated as C_2K_{12}. K_{12} can presumably be expressed as a function of temperature, and if necessary, of concentration. Experimentally, mixtures are equilibrated, concentrations measured and plotted as in Figure 5.1. Linear or nonlinear equations can then be fitted to the data. This approach is discussed in Section 5.6. The second approach involves the introduction of an intermediate quantity, a criterion of equilibrium, which can be related separately to C_1 and C_2. Chemical potential, fugacity, and activity are convenient criteria, with fugacity being preferred for most organic substances because of the simplicity of the equations which relate it to concentration. This approach is discussed in Section 5.6. The advantage of the equilibrium criterion approach is that properties of each phase are treated separately using a phase-specific equation. Treating phases in pairs, as is done with partition coefficients, can obscure the nature of the underlying phenomena. We may detect a variability in K_{12} and not know from which phase the variability derives. Furthermore, complications arise if we have 10 phases to consider. There are then 90 possible partition coefficients, of which only 9 are independent. Mistakes are less likely using an equilibrium criterion and the 10 equations relating it to concentration, one for each phase.

It is useful to discriminate between partition coefficients and distribution coefficients. A partition coefficient is strictly the ratio of the concentrations of the same chemical species in two phases. A distribution coefficient is a ratio of total concentrations of all species. Thus, if a chemical ionizes, the partition coefficient may apply to the unionized species, the distribution coefficient to ionized, and unionized species in total.

5.2 SOME THERMODYNAMIC FUNDAMENTALS

The Journey to Chemical Potential

There are four laws of thermodyamics numbered 0, 1, 2 and 3, because the need for the zeroth was not realized until after the first was discovered. Although these laws cannot be proved mathematically they are now universally accepted as true, or axiomatic, because they are supported by all available experimental evidence. On consideration, they are intuitively reasonable, and it now seems inconceivable that they are ever disobeyed.

The zeroth law introduces the concept of temperature as a criterion of thermal equilibrium by stating that when bodies are at thermal equilibrium, i.e., there is no net heat flow in either direction, their temperatures are equal.

The first law, which was discovered largely as a result of careful experiments by Joule, establishes the concept of energy and its conservation. Energy takes several forms—potential, kinetic, heat, chemical, electrical, nuclear, and electromagnetic. There are fixed conversion rates between these forms. Also, energy can neither be formed nor destroyed: it merely changes its form. Of particular importance are conversions between heat energy and work or mechanical energy.

The second law is intellectually more demanding, and introduces the concept of entropy and a series of useful related properties including chemical potential and fugacity. It is observed that whereas there are fixed exchange rates between heat and work energy, it is not always possible to effect the change. The conversion of mechanical energy to heat (as in an automobile brake) is always easy, but the reverse conversion of heat to mechanical energy (as in a thermal power station) proves to be more difficult. If a quantity of heat is available at high temperature, then only a fraction of it, perhaps one-third, can be converted to mechanical energy. The remainder is rejected as heat, but at a lower temperature. Most thermodynamics texts introduce hypothetical processes such as the Carnot cycle at this stage to illustrate these conversions. After some manipulation it can be shown that there is a property of a system termed entropy which controls these conversions. Apparently, regardless of how it is arranged to convert heat to work, the overall entropy of the system cannot decrease. It must increase by what is termed an irreversible process, or in the limit it could remain constant by what is called a reversible process. Although there may be a local entropy decrease, this must be offset by another and greater entropy increase elsewhere. Clausius summarized this law in the statement that the ''entropy of the universe increases.'' It can be shown that entropy is related to randomness or probability. An increase in entropy corresponds to a change to a more random, or disordered, or probable condition.

We are concerned with systems in which a chemical migrates from phase to phase. These phase changes involve input or output of energy, thus this energy change can compensate for entropy loss or gain. It can be shown that whereas entropy maximization is the criterion of equilibrium for a system containing constant energy at constant volume, the criterion at constant temperature and pressure (the environmentally relevant condition) is minimization of the related function, the Gibbs free energy G. G serves to combine energy and entropy in a common currency.

Returning to the example presented in Figure 5.1 earlier of benzene diffusing from water into an air bubble and striving to achieve equilibrium, the basic concept is that if we start with a benzene concentration in the water and none in air, the Gibbs free energy of the system will decrease as benzene migrates from water to air because the increase in free energy associated with the benzene concentration increase in the air is less than that of the decrease associated with benzene removal from the water. The process is thus spontaneous and irreversible. Benzene continues to diffuse from water into the air until it reaches a point at which the free energy increase in the air is exactly matched by the decrease in the water. At this point, the system comes to rest or equilibrium. Likewise if the system started with a higher benzene concentration in the air phase and approached equilibrium, it would reach exactly the same point of equilibrium with a particular ratio of concentrations in each phase.

The system thus seeks a minimum in G at which the derivative of G with respect to moles of benzene is equal in both air and water phases. This derivative is of such importance that it is given the symbol μ and is termed the chemical potential. The underlying principle of phase equilibrium thermodynamics is that when a solute such as benzene achieves equilibrium between phases such as air, water, and fish it seeks to establish an equal chemical potential in all phases. It will always diffuse from high to low chemical potential. Thus we can use chemical potential as a criterion of equilibrium for deductions of the direction of mass diffusion in the same way that we use temperature in heat transfer calculations.

Fugacity

Unfortunately, chemical potential is logarithmically related to concentration; thus, doubling a concentration does not double a chemical potential. A further complication is that chemical potentials cannot be measured absolutely; thus, it is necessary to establish some standard state at which it has a reference value. It was when addressing this problem that G. N. Lewis introduced a new function in 1901 which he termed 'fugacity', which has units of pressure and is assigned the symbol f. The term 'fugacity' comes from the Latin root fugere describing a 'fleeing' or 'escaping' tendency. It is identical to partial pressure in ideal gases and is logarithmically related to chemical potential. It is thus linearly or near-linearly related to concentration. Absolute values can be established because at low partial pressures under ideal conditions fugacity and partial pressure become equal. Thus we can replace the equilibrium criterion of chemical potential by that of fugacity. When benzene migrates between water and air it is seeking to establish an equal fugacity in both phases: its escaping tendency or partial pressures become equal in both phases.

Another useful quantity is the ratio of fugacity to some reference fugacity such as the vapor pressure of the pure chemical liquid. This is a dimensionless quantity and is termed "activity." Activity "a" can also be used as an equilibrium criterion. This proves to be preferable for substances such as ions, metals, or polymers which do not appreciably evaporate, and can thus not establish vapor phase concentrations and partial pressures.

Our task is then to start with a concentration of solute chemical in one phase; from this deduce the chemical potential, fugacity or activity; argue that these equilibrium criteria will be equal in the other phase; and then calculate the corresponding concentration in the second phase. We thus require recipes for deducing C from f and *vice versa*. This approach is depicted at the bottom of Figure 5.1.

The empirical or partition coefficient approach contains the inherent assumption that whatever the factors used to convert C_1 to fugacity and C_2 to fugacity their ratio is constant over a range of concentrations. Thus, it is not actually necessary to calculate the fugacities; their use is side-stepped. In the fugacity approach no such assumption is made and the individual calculations are undertaken. We can illustrate these approaches with an example.

Worked Example 5.1

Benzene is present in water at a specified temperature, and concentration C_1 of 1 mol/m³ (78 g/m³). What is the equilibrium concentration in air C_2?

(a) Fugacity approach.

Using techniques which we will devise later we find that for water under these conditions:

$$f_1 = C_1/0.002$$
$$= 500 \text{ Pa} = f_2 \text{ at equilibrium}$$
$$C_2 = 0.0004f_2 = 0.2 \text{ mol/m}^3 = 15.6 \text{ g/m}^3$$

(b) Partition coefficient approach.

K_{21} is 0.2, i.e. C_2/C_1
therefore $C_2 = K_{21}C_1 = 0.2 \times 1 = 0.2 \text{ mol/m}^3$ as above.

Clearly the problem is to find the correct conversion factors 0.002 and 0.0004, or K_{21}, which is their ratio. Care must be taken to avoid confusing K_{21} with its reciprocal K_{12} or C_1/C_2, which in this case has a value of 5.

The aims of this chapter are thus (a) to describe partition coefficients and present correlations or recipes for estimating them, (b) to justify and explain the use of fugacity (or in some cases activity) as a criterion of equilibrium, (c) to develop equations relating fugacity to concentration for each environmental phase, and (d) to suggest methods of obtaining data for application in these equations.

5.3 PARTITION COEFFICIENTS

5.3.1 Introduction

If we have two immiscible phases or media, for example air and water, or octanol and water, we can conduct experiments shaking volumes of both phases with a small amount of solute such as benzene to achieve equilibrium, then measure the concentrations and plot the results as was shown in Figure 5.1. It is preferable to use identical concentration units of amount per unit volume in each phase, but when one phase is solid it may be more convenient to express concentrations in units of amount per unit mass, such as $\mu g/g$, to avoid estimating phase densities. The data are often linear at low concentrations, thus we can write

$$C_1/C_2 = K_{12}$$

and the slope of the line is K_{12}.

We consider some nonlinear systems later.

The line may extend until some solubility limit or "saturation" is reached. In water this is the solubility of the chemical in water, which is usually available from handbooks. For many substances such as lower alcohols there is no "solubility" because the solute is miscible with water. In air the "solubility" is really the vapor pressure of the pure solute. This is the maximum partial pressure which the solute can achieve in the air phase. It is also available from handbooks. This vapor pressure (Pa) can be converted into a solubility with units of concentration by dividing by RT the gas constant, temperature product. Since C is n/V or moles divided by volume, and PV equals nRT

$$C \ (\text{mol/m}^3) = P(\text{Pa})/R(8.314 \ \text{Pa m}^3/\text{mol K}).T(K)$$

It is therefore convenient to estimate the air-water partition coefficient K_{AW} from

$$K_{AW} = C_A^S/C_W^S = P^S/RTC_W^S$$

where superscript S denotes saturation, P^S is the vapor pressure (Pa), and C_W^S is the solubility in water (mol/m^3).

Partition coefficients are widely available for systems of air-water, octanol-water, lipid-water, fat-water, hexane-water, "organic carbon"-water, various minerals-water, and activated carbon-water. These coefficients are related because they are really ratios of "solubilities" in one phase to that in water. We now discuss selected partition coefficients in more detail.

5.3.2 Air-Water Partition Coefficients

The nature of air-water partition coefficients or Henry's law constants has been reviewed by Mackay and Shiu (1981). Only a brief summary is given here. As was discussed above, the simplest method of estimating Henry's law constants of organic solutes is as a ratio of vapor pressure to water solubility. It must be recognized that this contains the inherent assumption that water is not very soluble in the organic material because the vapor pressure which is used is that of the pure substance (normally the pure liquid), whereas in the case of solubility of a liquid such as benzene in water, the solubility is not actually that of pure benzene, but is inevitably that of benzene saturated with water. When the solubility of water in a liquid exceeds a few percent, this assumption may break down and it is unwise to use this relationship. If a solute is miscible with water (e.g., ethanol), it is preferable to determine the Henry's law constant directly; that is, by measuring air and water concentrations at equilibrium. This can be done by various techniques, e.g., the EPICS method described by Gossett (1987); a continuous stripping technique described by Mackay et al. (1979); or a "wetted wall" technique devised by Fendinger and Glotfelty (1988). A desirable strategy is to measure vapor pressure P^S, solubility C^S, and H or K_{AW} and perform an internal consistency check that H is indeed P^S/C^S, or close to it.

Care must be taken when calculating Henry's law constants to ensure that the vapor pressures and solubilities apply to the same temperature, and to the same phase. In some cases reported vapor pressures are estimated by extrapolation from higher temperatures. They may be of a liquid or subcooled liquid, whereas the solubility is that of a solid. Figure 5.2 is the familiar P^S,T phase diagram which illustrates the dependence of P^S on T for solid, liquid, and subcooled liquid states. Subcooled conditions are not experimentally accessible, but prove to be useful for theoretical purposes, as is discussed later.

Henry's law constants vary over many orders of magnitude, tending to be high for substances such as the alkanes which have high vapor pressures, low boiling points, and low solubilities, and very low for substances such as alcohols which have a high solubility in water and a low vapor pressure. There is a common misconception that substances which are "involatile", such as DDT, will have a low Henry's law constant. This is not necessarily the case because these substances also have very low solubilities in the water, i.e., they are very hydrophobic; thus their low vapor pressure is offset by their low water solubility, and they have relatively large Henry's law constants. They may thus partition appreciably from water into the atmosphere and evaporate from rivers and lakes.

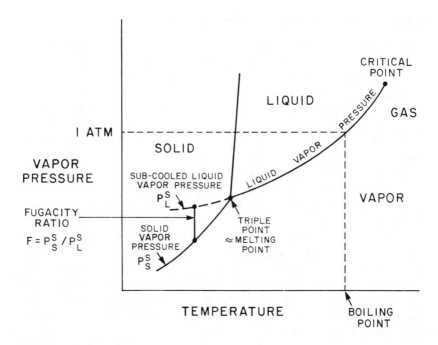

Figure 5.2.   PT diagram for a pure substance.

Figure 5.3 is a plot of P^S versus C^S on which chemicals of constant H lie on the 45° diagonals. It is noteworthy that some quite involatile chemicals have similar H's to volatile chemicals because of the compensating solubility. A homologous series of chemicals, such as the chlorobenzenes or PCBs, tend to lie close to a diagonal with similar H values.

Worked Example 5.2

Deduce H and K_{AW} for benzene and DDT given the following data at 25°C.

	Mol. Mass (g/mol)	Solubility (g/m³)	Vapor Pressure (Pa)
benzene	78	1780	12700
DDT	354.5	.003	0.00002
2,4-D	221	890	0.000056

In each case the solubility C^S in mol/m³ is the solubility in g/m³ divided by the molecular mass, e.g., 1780/78 or 22.8 mol/m³ for benzene. H is then P^S/C^S or 556 Pa m³/mol for benzene. K_{AW} is H/RT or 556/(8.314 × 298) or 0.22.

Similarly for DDT, H is 2.36 and K_{AW} is 9.5×10^{-4}. For 2,4-D, H is 1.39 $\times 10^{-5}$ and K_{AW} is 5.6×10^{-9}.

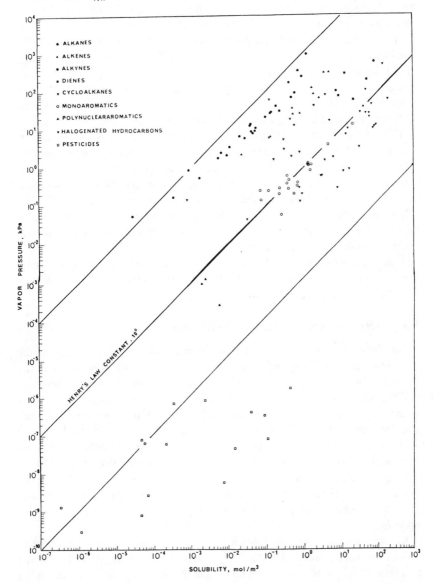

Figure 5.3.   Plot of P^S versus C^S for selected chemicals, showing diagonals of constant H.

Note that DDT and 2,4-D (a common herbicide) have similar vapor pressures but very different H and K_{AW} values because of their solubility differences. Likewise, the vapor pressure of DDT is about 600 million times less than that of benzene but H is only 200 times less, because of DDT's very low solubility. Benzene tends to evaporate appreciably from water into air, DDT less so but still to a significant extent, while 2,4-D does not evaporate to any significant extent.

A complication encountered in experimental measurement of H is the tendency for hydrophobic chemicals to associate with the air-water interface. This has been discussed by Valsaraj (1988) and is the basis of the "solvent sublation" separation process (Valsaraj et al., 1986). Since this is an area effect, it is observed most readily when the area/volume ratio is large, which implies small air bubbles or water drops. Glotfelty et al. (1987) have observed that some pesticides are enriched in fog droplets at concentrations several thousand times that expected from H. This may be a surface effect.

H is also affected by any factor which affects solubility in water, such as the presence of electrolytes or sorbents. H is also very temperature-sensitive.

Compilations of H and K_{AW} data, or discussions of estimation methods can be found in Mackay and Shiu (1981), Shiu and Mackay (1986) for PCBs, Suntio et al. (1988) for pesticides, Ashworth et al. (1988), Nirmalakhandan and Speece (1988b), and Lyman et al. (1982).

5.3.3   Octanol-Water Partition Coefficients

The dimensionless octanol-water partition coefficient (K_{OW}) is one of the most important and frequently used descriptors of chemical behavior in the environment. The use of 1-octanol has been popularized by Hansch and Leo (1979), who have sought correlations between K_{OW} and many biochemical phenomena, and have compiled extensive data bases. Various methods are available for calculating K_{OW} from molecular structure, as reviewed by Lyman et al. (1982). Octanol was selected because it has a similar carbon to oxygen ratio as lipids, is readily available in pure form, and is only sparingly soluble in water (0.58 g/L). Water is, however, quite soluble in octanol (41 g/L) (Lyman et al., 1982). In the pharmaceutical and biological literature K_{OW} is given the symbol P (for partition coefficient), which we prefer to reserve for pressure.

K_{OW} is a measure of hydrophobicity, i.e., the tendency of a chemical to "hate," or partition out of water. As was discussed earlier, it can be viewed as a ratio of solubilities in octanol and water, but in most cases of liquid chemicals there is no actual, measurable solubility in octanol because octanol and the liquid are miscible. The "solubility" of organic chemicals in octanol tends to be fairly constant in the range 200 to 2000 mol/m^3, thus the variation in K_{OW} between chemicals is primarily due to the variation in water solubility. It is thus misleading to assert that K_{OW} describes lipophilicity or "love for lipids" because most chemicals "love" lipids equally, but they "hate" water quite differently.

Because K_{OW} varies so greatly from approximately 0.1 to 10^7 it is common to express it as log K_{OW}. It is a disastrous mistake to use log K_{OW} in a calculation when K_{OW} should be used! Compilations of K_{OW} values and calculation methods have been prepared by Hansch and Leo (1979), Lyman et al. (1982), Kier and Hall (1977, 1986), Rekker (1977) and Sangster (1989).

We treat the thermodynamics of K_{OW} and its relationship to solubility later in this chapter.

5.3.4 Organic Carbon-Water Partition Coefficients

Studies by agricultural chemists have revealed that hydrophobic organic chemicals tended to sorb primarily to the organic carbon present in soils. Similar observations have been made for bottom sediments. In a definitive study Karickhoff (1981) showed that it was organic carbon which was almost entirely responsible for the sorbing capacity of sediments, and that the partition coefficient between sediment and water expressed in terms of an organic carbon partition coefficient (K_{OC}) was closely related to the octanol-water partition coefficient. Indeed, the simple relationship was established

$$K_{OC} = 0.41 \ K_{OW}$$

This relationship is based on experiments in which a soil-water partition coefficient was measured for a variety of soils of varying organic carbon content (y) and chemicals of varying K_{OW}. The concentration in soil was measured in units of μg/g or mg/kg (usually of dry soil), and the water in units of μg/cm^3 or mg/L. The ratio of soil and water concentration (designated K_P) thus has units of L/kg or reciprocal density.

$$K_P = C_S/C_W = (mg/kg)/(mg/L) = L/kg$$

If a truly dimensionless partition coefficient is desired it is necessary to multiply K_P by the soil density in kg/L (typically 2.5) or equivalently, multiply C_S by density to give a concentration in units of mg/L.

A plot of K_P versus organic carbon content y(g/g) for a variety of soils proves to be near-linear and passes close to the origin, suggesting the relationship

$$K_P = y \ K_{OC}$$

where K_{OC} is an organic carbon to water partition coefficient.

In practice there is usually a slight intercept; thus, the relationship must be used with caution when y is less than 0.01, and especially less than 0.001. Since y is dimensionless K_{OC}, like K_P, has units of L/kg.

Measurements of K_{OC} for a variety of chemicals have shown that K_{OC} is related to K_{OW}, as discussed above. K_{OW} is dimensionless, thus the constant 0.41 has dimensions of L/kg.

Care must be taken in these calculations to use consistent units. For example, if the water concentration has units of mol/m^3, and K_P is applied, the soil concentration will be in mol/Mg, i.e., moles per 10^6 grams. The usual units used are mg/L in water and mg/kg in soil. Any unit can be used for the amount of solute, i.e., g or mol, but it must be consistent in both water and soil.

These relationships provide a very convenient method of calculating the extent of sorption of chemicals between soils or sediments and water, provided that the organic carbon content of the soil and the chemicals' octanol-water partition coefficient are known. This is illustrated in Example 5.3 below.

Worked Example 5.3

Estimate the partition coefficient between a soil containing 2% (mass) of organic carbon for benzene (K_{OW} of 135) and DDT (log K_{OW} of 6.19), and the concentrations in soil in equilibrium with water containing .001 g/m^3 of each chemical.

benzene	$K_{OC} = 0.41\ K_{OW} = 55,$	$K_P = 0.02\ K_{OC} = 1.1$
DDT	$K_{OC} = 0.41\ K_{OW} = 635000,$	$K_P = 0.02\ K_{OC} = 12700$

K_P and K_{OC} have units of L/kg or m^3/Mg, i.e., reciprocal density; thus, when applying the equation below, C_S, the soil concentration, will have units of mg/kg or μg/g

$C_S = K_P C_W$	benzene	$C_S = 1.1 \times 0.001 = 0.0011$ mg/kg
	DDT	$C_S = 12700 \times 0.001 = 12.7$ mg/kg

Note the much higher DDT concentration in the soil because of its hydrophobic character.

The concentrations in the organic carbon are $C_W K_{OC}$ or .055 μg/g for benzene and 635 μg/g for DDT. If octanol was exposed to this water, similar concentrations of $C_W K_{OW}$ or 0.135 g/m^3 or μg/cm^3 for benzene and 1550 μg/cm^3 for DDT would be established in the octanol.

The relationship between K_{OW} and K_{OC} has been the subject of considerable investigation and it appears to be somewhat variable; for example, DiToro (1985) has suggested that for suspended matter in water, K_{OC} approximately equals K_{OW}. Other workers, notably Gauthier et al. (1987) have shown that the sorbing quality of the organic carbon varies and appears to be related to its aromatic content, as revealed by NMR analysis. The text edited by Suffett and McCarthy (1989) discusses this issue in some detail. We thus conclude that K_{OC} and K_{OW} are closely related, but using Karickhoff's relationship may involve an error of a factor of two in either direction.

5.3.5 Lipid-Water Partition Coefficients

Studies of fish-water partitioning by workers such as Spacie and Hamelink (1982), Neely (1979), Veith et al. (1979), and Mackay (1982) have shown that the primary sorbing or dissolving medium in fish for hydrophobic organic chemicals is lipid or fat. A similar approach can be taken as for soils, but in this case there is a more reliable relationship between K_{OW} and K_{LW}, the lipid-water partition coefficient. For most purposes they can be assumed to be equal, although for the very hydrophobic substances Gobas et al. (1988a) suggest that this breaks down, possibly because of the structured nature of the lipid phases. It is thus possible to calculate an approximate fish to water bioconcentration factor or partition coefficient if the lipid content of the fish is known. Mackay (1982) reanalyzed a considerable set of bioconcentration data and suggested the simple linear relationship

$$K_{FW} = 0.048 \, K_{OW}$$

This can be viewed as expressing the assumption that fish is about 5% lipid. Lipid contents vary considerably, and it is certain that there is some sorption to non-lipid material, but it appears that on the average the fish behaves as if it is about 5% octanol by volume.

This relationship should not be used for very small organisms such as plankton which have large area/volume ratios and may have higher apparent partition coefficients because of surface sorption.

5.3.6 Mineral Matter-Water Partition Coefficients

Partition coefficients of hydrophobic organics between mineral matter and water are generally fairly low and do not appear to be simply related to K_{OW}. Typical values of the order of unity to 10 are observed. A notable exception occurs when the mineral surface is dry. Dry clays display very high sorptive capacities for organics, probably because of the activity of the inorganic sorbing sites. This raises a problem that some soils may display highly variable sorptive capacities as they change water content as a result of heating, cooling, and rainfall during the course of diurnal or seasonal variations. Some pesticides are supplied commercially in the form of the active ingredient sorbed to an inorganic clay such as bentonite.

In the environment, most clay surfaces appear to be wet and thus of low sorptive capacity. Most mineral surfaces which are accessible to the biosphere also appear to be coated with organic matter, probably of bacterial origin. They may thus be shielded from the solute by a layer of highly sorbing organic material. It is thus a fair (and very convenient) assumption that the sorptive capacity of clays and other mineral surfaces can be ignored. Notable exceptions to this are subsurface (groundwater) environments in which there may be extremely low organic carbon contents, and sorption of certain ionizing solutes such as phenols or acids. In such cases the inherent sorptive capacity of the mineral matter may be controlling.

5.3.7   Aerosol-Air Partition Coefficients

One of the most difficult, and to some extent puzzling, sorption partition coefficients is that between air and aerosol particles. These particles have very high specific areas, i.e., areas per unit volume. They also are found to be very effective sorbents. It is not usual to calculate the aerosol to air partition coefficient; rather, partitioning is commonly expressed in terms of the fraction of the chemical which is in particulate or gaseous form. This is normally measured experimentally by passing a volume of air through a filter, then measuring the concentrations in air before and after filtration, and of the trapped particles. Relationships can then be established between the ratio of gaseous to aerosol material and the concentration of total suspended particulates (TSP).

For example, Yamasaki et al. (1982) have measured gaseous (A) and particle associated or filterable concentrations (F) (per volume of air) for PAHs over a range of temperatures and have developed correlations of the form

$$\log (A.TSP/F) = B_1 - B_2/T$$

where B_1 and B_2 are constants specific to the hydrocarbon and TSP is total suspended particulate concentration. Bidleman et al. (1986), Ligocki and Pankow (1989) and Foreman and Bidleman (1987) have generated and correlated similar data. Pankow (1987, 1988) and Bidleman (1988) have discussed these and other results and underlying theoretical principles, and especially the dependence of the A.TSP/F group on the chemical's vapor pressure.

Essentially, the group A.TSP/F is a partition coefficient in disguise because (F/TSP) is the concentration on the particles, e.g., $\mu g/g$, F being $\mu g/m^3$ air and TSP being g particle/m^3 air, or related units. The constant B_2 is related to the enthalpy of sorption.

A simple procedure has been suggested by Mackay et al. (1986) that the dimensionless partition coefficient K_{XA} between aerosol (X) and air (A) is given by the relationship below

$$K_{XA} = 6 \times 10^6/P_L^S$$

where P_L^S is the vapor pressure of the subcooled liquid solute. It is thus apparent that substances of very low vapor pressure, i.e., those which are very "insoluble" in air, will have high partition coefficients to aerosol particles and thus become subject to wet and dry deposition processes. Notable among these chemicals are the high molecular weight hydrocarbons including the polynuclear aromatic hydrocarbons, the higher PCBs, and chlorinated dioxins and dibenzofurans. They are often referred to as "semivolatile" chemicals. Involatile metals are of course almost totally particle-associated.

5.4 SORPTION BY DISPERSED PHASES

A frequent environmental calculation is to estimate the fraction of a chemical that is present in a fluid which is sorbed to some dispersed sorbing phases within that fluid. An example is the fraction of material attached to aerosols or associated with suspended solids or with biotic matter including fish. The reason for this calculation is that often the measured concentration is of the total (i.e., dissolved and sorbed) chemical and it is useful to know what fractions are in each phase. This is particularly useful when subsequently calculating rates of evaporation or uptake of chemical by fish from water, in which the partitioning may be only from the dissolved solute.

To establish the general equations describing sorption in such cases we designate the continuous phase by subscript A and the dispersed phase by subscript B. The dispersed phase volume is typically a factor of 10^{-5} or less, than that of the continuous phase.

The volumes (m³) are V_A and V_B, and usually, V_A greatly
    exceeds V_B
The equilibrium concentrations are C_A and C_B mol/m³
The dimensionless partition coefficient is $K_{BA} = C_B/C_A$
The total amount of solute is M moles.

$$M = V_A C_A + V_B C_B = V_T C_T$$

where C_T is the total concentration. It can be assumed that V_T the total volume is approximately V_A. Now

$$C_B = K_{BA} C_A$$

therefore $M = C_A(V_A + V_B K_{BA}) = C_T V_A$
from which $C_A = C_T/(1 + K_{BA} V_B/V_A) = C_T/(1 + K_{BA} x_B)$

where x_B is the volume fraction of phase B, i.e., V_B/V_A.

The fraction dissolved (i.e., in the continuous phase) is

$$C_A/C_T = 1/(1 + K_{BA}X_B)$$

and that sorbed is

$$(1 - C_A/C_T) = K_{BA}X_B/(1 + K_{BA}X_B)$$

The key quantity is thus $K_{BA}X_B$ or the product of the dimensionless partition coefficient and the volume fraction of the dispersed sorbing phase. When this product is 1.0, half the solute is in each state. When it is smaller than 1.0, most is dissolved, and when it exceeds 1.0 more is sorbed.

When the phase B is solid, it is usual to express the concentration C_B in units of moles or grams per unit mass of B, in which case K_{BA} has units of volume/mass or reciprocal density. For example, it is common to use mg/L for C_A, mg/kg for C_B, and L/kg for K_P, then with total mass M in mg, and V_A in L, it can be shown that

$$M = C_A(V_A + m_BK_P) = C_TV_A$$

where m_B is the mass of sorbing phase (kg), from which

$$C_A = C_T/(1 + K_Pm_B/V_A) = C_T/(1 + K_PX_B)$$

where X_B is the concentration of sorbent in kg/L. The units of the partition coefficient K_{BA} or K_P and concentration of sorbent x_B or X_B do not matter as long as their product is dimensionless and consistent, i.e., the amounts of sorbing phase, continuous phase, and chemical are the same in the definition of both the partition coefficient and the sorbent concentration.

Distribution coefficients usually are ratios of concentrations in a phase to the concentration C_T, whereas partition coefficients are ratios to C_A, the smaller dissolved value.

Worked Example 5.4

An aquarium contains 10 m^3 of water and 200 fish, each of volume 1 cm^3. How will 0.01 g, i.e., 10 mg, of benzene and DDT partition between water and fish, given that the fish are 5% lipids and log K_{OW} values are 2.13 (benzene) and 6.19 (DDT)?

> K_{FW} will be 0.05 K_{OW} or 6.7 for benzene and 77400 for DDT.
> C_T is 0.001 g/m^3 in both cases.

Fraction dissolved is

$$1/(1 + K_{FW}x_F)$$

where x_F is the volume fraction of fish, i.e., $200 \times 10^{-6}/10 = 2 \times 10^{-5}$.

For benzene, $K_{FW}x_F$ is 0.00013
For DDT, $K_{FW}x_F$ is 1.55

Therefore, the fractions dissolved are 0.99987 for benzene and 0.39 for DDT. The dissolved concentrations are thus 0.00099987 g/m^3 (benzene) and 0.00039 g/m^3 (DDT), and the sorbed concentrations (per m^3 of water) are 0.00000013 g/m^3 and 0.00061 g/m^3, respectively. The sorbed concentrations per m^3 of fish are 0.0067 g/m^3 and 30 g/m^3, respectively. Note the appreciable bioconcentration of DDT.

Example 5.5

A lake of volume 10^6 m^3 contains 15 mg/L of sorbing material. The total concentration of a chemical of K_P equal to 10^5 L/kg is 1 mg/L. What are the dissolved and sorbed concentrations and amounts?

Answer: 0.4 mg/L, or 400 kg dissolved and 0.6 mg/L or 600 kg sorbed.

5.5 MULTIMEDIA PARTITION COEFFICIENT CALCULATIONS

There is no doubt that the calculation of one phase concentration from another through the use of a simple partition coefficient is the most direct and convenient method. Care must be taken that the concentration units and the partition coefficient dimensions are consistent, especially when dealing with solid phases. There may also be inadvertent inversion of a partition coefficient, i.e., the use of K_{12} instead of K_{21}. It is also possible to deduce certain partition coefficients from others, e.g., if an air/water and a soil/water partition coefficient are available, then the air/soil or soil/air partition coefficient can be deduced as follows:

$$K_{AS} = K_{AW}/K_{SW}$$

If we are treating 10 phases, it is possible to define 9 independent interphase partition coefficients, the tenth being dependent on the other 9. In principle, with 10 phases it is possible to define 90 partition coefficients, half of which are reciprocals of the others. When dealing with very complex multicompartment environmental media, extreme care must be taken to avoid overspecifying or under-specifying partition coefficients and ensuring that the ratios are consistent and not inverted.

The primary difficulty with partition coefficients is that each numerical value depends on the properties of the chemical in two phases; thus, it is not immediately clear what is controlling the absolute magnitude of partition coefficients. For example, if two hydrocarbons have octanol/water partition coefficients of 100 and 1000, is this because they behave 10 times differently in water, or 10 times differently in the octanol, or some combination of these two? If we wish to develop reliable thermodynamically based correlations for partition coefficients, we have to probe more deeply into their origin and establish what is controlling them. This is treated in Section 5.6.

We have, however, developed the capability of performing our first multimedia partitioning calculations. If we have a series of phases of, for example, volume V_1, V_2, V_3, and V_4, and we know the partition coefficients K_{21}, K_{31}, K_{41} and we introduce a known amount (M) of chemical into this hypothetical environment, then we can argue, from the mass balance principle, that M will be the sum of the concentration-volume products as follows:

$$
\begin{aligned}
M &= C_1V_1 + C_2V_2 + C_3V_3 + C_4V_4 \\
&= C_1V_1 + (K_{21}C_1)V_2 + (K_{31}C_1)V_3 + (K_{41}C_1)V_4 \\
&= C_1[V_1 + K_{21}V_2 + K_{31}V_3 + K_{41}V_4]
\end{aligned}
$$

therefore

$$C_1 = M/[V_1 + K_{21}V_2 + K_{31}V_3 + K_{41}V_4]$$

and $C_2 = K_{21}C_1$ etc. and amount $m_1 = C_1V_1$, etc.

It is thus possible to calculate the concentrations in each phase, the amounts in each phase, and the percentages, and obtain a tentative picture of the behavior of this chemical at equilibrium in an evaluative environment. This is best illustrated by an example.

Worked Example 5.6

Benzene partitions in a hypothetical environment between air, water, sediment, and fish (subscripted A, W, S, and F). The volumes of each phase being given below. The dimensionless partition coefficients are also given below. Calculate the concentrations, amounts, and percentages in each phase, assuming that a total of 10 moles of benzene is introduced into this system.

$$
\begin{array}{llll}
V_A = 1000 & V_W = 20 & V_S = 10 & V_F = 0.05 \\
K_{AW} = 0.2 & K_{SW} = 15 & K_{FW} = 20 &
\end{array}
$$

It follows, using the equation developed above, that

$$C_W = 10/(20 + 0.2 \times 1000 + 15 \times 10 + 20 \times 0.05) = 10/371 = 0.027$$

therefore

$$C_A = 0.0054, \; C_S = 0.405, \; C_F = 0.54$$

The amounts in each phase are the products CV, namely

air = 5.39, water = 0.54, sediment = 4.04, fish = 0.03

from which the percentages are, respectively, 53.9, 5.4, 40.4, and 0.3.

It is clear that, in this system, benzene partitions primarily into air, largely because of the large air volume. The concentration in the air is lower than in any other phase; thus, we must be careful to discriminate between where the *amount* of chemical is large, which depends on the product CV, and where the *concentration* is large. There is a high concentration in the fish, but only a negligible percentage of the benzene is associated with fish. Such calculations are invaluable because they establish the predominant medium in which the chemical is likely to partition, and they even give approximate relative concentrations. McCall et al. (1983) have, for example, successfully estimated pesticide partitioning in soils using this simple approach.

Example 5.7

Repeat the above example for DDT and p-cresol using the properties given earlier in Table 3.2.

5.6 RELATING CONCENTRATION AND FUGACITY: Z VALUES

5.6.1 The Definition of Z

The material in this section is taken from a series of publications introducing and discussing the fugacity concept, notably Mackay (1979), Mackay and Paterson (1981, 1982), Paterson and Mackay (1985) and Paterson (1985).

We seek working equations relating fugacity (f) to concentration (C) for each chemical and each environmental medium. The simplest approach is to exploit the expected near-linear relationship between f and C by postulating

$$C = Zf$$

where Z is a proportionality constant, termed the "fugacity capacity," with units of mol/m^{3}Pa. This equation does not necessarily imply that C and f are always linearly related. Non-linearity can be accommodated by allowing Z to vary as a function of C or f.

We expect that Z will depend on

(a) the nature of the solute (chemical)
(b) the nature of the medium or compartment
(c) temperature

(d) pressure (but the effect is usually negligible)
(e) concentration (but the effect is usually negligible at low concentrations)

If we can develop procedures by which Z values can be estimated for any given environmental situation, equilibrium concentrations can then be deduced using f as a common criterion of equilibrium.

The fugacity capacity Z may at first be a difficult concept to grasp because it has unfamiliar units of mol/(volume × pressure). Heat capacity calculations provide a precedent for introducing Z and may help to illustrate the fundamental nature of this quantity.

5.6.2   The Heat Capacity Analogy to Z

The traditional heat capacity equation is written in the form

$$\text{Heat Content (J)} = \text{Mass of phase (kg)} \times$$
$$\text{Heat Capacity (J/kg.K)} \times \text{Temperature (K)}$$

For example, water has a heat capacity of 4180 J/kg.K, which is more familiar as 1 cal/g.°C. We can rearrange this equation to give

$$\text{Heat Concentration (J/m}^3) = \text{Heat Capacity (J/m}^3\text{K)} \times \text{Temperature (K)}$$

This new volumetric heat capacity for water is 4180000 J/m³K. The use of mass rather than volume in heat capacities is an ''accident'' resulting from the greater ease and accuracy of mass measurements compared to volume measurements, and the complication which arises that volumes change on heating, while mass remains constant.

The equilibrium criterion used above is temperature (K), whereas we are concerned with fugacity (Pa). The quantity which partitions above is heat (J), whereas we are concerned with amount of matter (moles). Replacing K by Pa and J by mol leads to the fugacity equation

$$C \text{ (mol/m}^3) = Z \text{ (mol/m}^3\text{Pa)} \times f \text{ (Pa)}$$

Z is thus analogous to a heat capacity.

Experience with heat calculations leads to a mental concept of heat capacity as a property describing the "capacity of a phase to absorb heat for a certain temperature rise." For example, if 1 g of water (heat capacity 4.2 J/g°C) absorbs 4.2 J, its temperature will rise 1°C. Copper with a lower heat capacity of 0.38 J/g°C requires absorption of only 0.38 J to cause the same rise in temperature. Hydrogen gas has a large heat capacity of 14.3 J/g°C and thus requires a great deal of heat to raise its temperature. These substances differ markedly in their temperature response when heat is added. Hydrogen has a very large capacity for heat, water a large capacity, copper a low capacity. If 1000 J are added to equal masses of 1 g of these substances, the copper becomes much hotter by 263°C (or 100/.38), while the water only heats up by 24°C (or 100/4.2), and the hydrogen by only 7°C (or 100/14.3). Hydrogen and water can thus absorb or "soak up" larger quantities of heat without becoming much hotter.

The fugacity capacity is similar. Phases of high Z (possibly sediments or fish) are able to absorb a greater quantity of solute, yet retain a low fugacity. It follows that solutes will tend to partition into these high Z phases and build up a substantial concentration, yet retain a relatively low fugacity. Conversely, phases of low Z will tend to experience a large fugacity increase following absorption of a small quantity of solute. A substance such as DDT is readily absorbed by fish and achieves high concentration while maintaining a low fugacity. The Z value of DDT in fish is large. On the other hand, DDT is not readily absorbed by water; indeed, DDT is hydrophobic or "water hating." Its Z value in water is very low.

This analogy between heat and fugacity capacity is perhaps best illustrated by the following pair of numerically identical examples; the fugacity quantities being given in parentheses.

Worked Example 5.8

A system consists of three phases 10 g of water (10m^3 of water) of heat capacity 4.2 J/g°C (fugacity capacity 4.2 mol/m^3 Pa), 5 g of copper (5 m^3 of air) of heat capacity 0.38 J/g°C (fugacity capacity 0.38 mol/m^{3}Pa), and 1 g of hydrogen (1m^3 of sediment) of heat capacity 14.3 J/g°C (mol/m^3 Pa). To this system is added 582 J of heat (582 mols of solute). What is the heat (solute) distribution at equilibrium and what is the rise in temperature (fugacity) and heat concentrations in J/g (molar concentrations in mol/m^3)? We assume, for simplicity, that the initial temperature is 0°C and the initial concentrations are also zero. (Note that Z for a solute in a gas never has the above value.)

Solution

In approaching equilibrium the temperatures (fugacities) will rise equally to a new common value at T°C, (f Pa) such that the amount of heat (solute) in each phase will be

Amount of Heat (J) = Mass (g) × Heat Capacity (J/g°C) ×
Temperature (°C) or
Amount of Matter (mol) = Volume (m^3) × Fugacity Capacity
(mol/m^{3}Pa) × Fugacity (Pa)

Thus the total will be the summation over the 3 phases

i.e., $582 = 10 \times 4.2 \times T + 5 \times 0.38 \times T + 1 \times 14.3 \times T$

thus

$T = 582/(10 \times 4.2 + 5 \times 0.38 + 1 \times 14.3) = 10\,°C\ (Pa)$

1. Heat (moles) in water $= 10 \times 4.2 \times 10$ $\qquad = 420J$ (moles) 72%
2. Heat (moles) in copper (air) $= 5 \times 0.38 \times 10 = 19J$ (moles) 3%
3. Heat (moles) in hydrogen (sediment) $= 1 \times 14.3 \times 10$

$\qquad\qquad\qquad\qquad\qquad\qquad\qquad = 143J$ (moles) 25%

$\qquad\qquad\qquad\qquad\qquad$ Total $\qquad 582J$ (moles)100%

The concentrations are

Water	$4.2 \times 10 = 42$ J/g (mol/m³)
Copper (air)	$0.38 \times 10 = 3.8$ J/g (mol/m³)
Hydrogen (sediment)	$14.3 \times 10 = 143$ J/g (mol/m³)

The distribution of heat (moles) is influenced by the relative phase masses (volumes) and the heat capacities (Z values). Despite the fact that the third phase is small in volume, its much larger heat capacity (Z) results in accumulation of a substantial fraction of the total (25%) and its concentration is a factor of 3.4 and 38 greater than the other two phases—which is, of course, the ratio of the heat capacities (or ratio of Z values, this ratio being the partition coefficient).

This example could have been solved using heat capacity partition coefficients, but of course no such quantities are tabulated in handbooks. Indeed, any suggestion that heat partition coefficients are useful would be treated with derision. In environmental calculations, on the other hand, the use of Z values is more novel and the use of partition coefficients is routine. In essence, the use of fugacity capacities is an attempt to bring to environmental calculations some of the procedural benefits which are routinely enjoyed by the use of heat capacities.

It transpires that this illustrative example is a "Level I" fugacity calculation as described later and is the simplest possible environmental equilibrium model calculation. It is ultimately algebraically identical to the partitioning example 5.6.

Worked Example 5.9

A three phase system has Z values $Z_1 = 5 \times 10^{-4}$, $Z_2 = 1.0$, $Z_3 = 20$ (all mol/m³ Pa) and volumes $V_1 = 1000$, $V_2 = 10$ and $V_3 = 0.1$ (all m³). Calculate the distributions, concentrations, and fugacity when 1 mol of solute is distributed at equilibrium between these phases. It is suggested that the calculations be done in tabular form.

Phase	Z	V	VZ	C = Zf	VC	%
1	5×10^{-4}	1000	0.5	4×10^{-5}	0.04	4
2	1.0	10	10	0.08	0.80	80
3	20	0.1	2	1.6	0.16	16
Total			12.5		1.0	100

$$M = V_1 Z_1 f + V_2 Z_2 f + V_3 Z_3 f = f \Sigma VZ$$

therefore

$$f = M / \Sigma VZ = 1.0/12.5 = 0.08$$

Again, a large value of Z or C does not necessarily imply a large quantity.

5.6.3 Methods of Estimating Z Values

There are two procedures for defining Z for a solute in a medium. The first is to write an appropriate thermodynamic equation for fugacity and manipulate it into a form so that a group of variables corresponding to Z is defined, namely

$$C = (Group) f$$

This is satisfactory for air and water because we can use already established thermodynamic equations.

For other phases such as fish, soils, sediments, and aerosols we adopt another procedure exploiting partition coefficients. If we know Z for phase 1 (say, solution in water) and we can determine a dimensionless partition coefficient K_{21} (e.g., fish-water) which is C_2/C_1, then since the fugacities of the solute are equal at equilibrium

$$K_{21} = C_2/C_1 = Z_2 f/Z_1 f = Z_2/Z_1$$

therefore

$$Z_2 = Z_1 K_{21}$$

We can thus deduce Z_2, i.e., the fugacity capacity in fish. We can "hop" from medium to medium calculating a new Z at each "hop." The Z values actually start in the gas or air phase and "hop" to water, and from there to sediments, soil, or fish. Later we encounter cases in which this is inconvenient and it is preferable to start in the water phase.

5.6.4   Z for Air

The basic fugacity equation for a chemical in the vapor state as presented in thermodynamics texts (Prausnitz et al., 1986) is

$$f = y \, \phi \, P_T$$

where y is mole fraction of chemical, ϕ is a fugacity coefficient, and P_T is total (atmospheric) pressure. If the gas law applies

$$P_T V = nRT$$

where n is the total number of moles of air and chemical present, R is the gas constant (8.314 Pa m³/mol K), and T is absolute temperature. Now the concentration of the chemical or solute C will be yn/V, thus

$$C = yP_T/RT = (1/\phi RT) \, f$$

thus Z is $1/\phi RT$ or $1/RT$ when ϕ is 1.0. Fortunately the fugacity coefficient ϕ rarely deviates appreciably from unity under environmental conditions. The exceptions occur at low temperatures, high pressures, or when the solute molecules interact chemically with each other in the gas phase. Only this last class is important environmentally. Carboxylic acids such as formic and acetic acid tend to dimerize, as do certain gases such as NO_2.

It is noteworthy that Z for air is thus essentially identical for all non-interacting solutes at the same temperature, having a value of approximately 4.1×10^{-4} mol/m³ Pa at environmental conditions. It has an obvious temperature dependence and no significant pressure dependence. The fugacity is then numerically equal to the partial pressure of the solute. Concentrations are obtained from partial pressures by dividing by RT.

This raises a question as to why we use the term "fugacity" in preference to "partial pressure." The answers are that (a) under conditions when ϕ is not unity, fugacity and partial pressure are not equal, and (b) there is some conceptual difficulty about referring to a "partial pressure of DDT in a fish" when there is no vapor present for a pressure to be present in—even partially.

5.6.5   Z for Water

The basic fugacity equation (Prausnitz et al., 1986) for a chemical i dissolved in water, or indeed any solvent, is given in terms of mole fraction x_i, activity coefficient γ_i and reference fugacity f_R, on a Raoult's law basis as

$$f_i = x_i \gamma_i f_R$$

Now x_i the mole fraction of solute chemical can be converted to concentration C mol/m³ using molar volumes v (m³/mol), amounts n (mol), and volumes V (m³) of solute chemical (subscript i) and water (subscript w). Assuming that the solute concentration is small we can write

$$C_i = n_i/(V_w + V_i) \approx n_i/V_w$$

But
$$V_w = n_w v_w$$

and
$$x_i = n_i/(n_i + n_w) \approx n_i/n_w$$

thus
$$C_i \approx x_i/v_w \approx x_i/(18 \times 10^{-6} \text{ m}^3/\text{mol})$$

thus
$$f_i = C_i v_w \gamma_i f_R$$

or
$$C_i = (1/v_w \gamma_i f_R) f_i$$

thus
$$Z_i \text{ is } 1/v_w \gamma_i f_R.$$

The reference fugacity f_R is by definition the fugacity which the solute will have (or tend to) when in the pure liquid state when x_i is 1.0, and γ_i is also (by definition) 1.0. This is then the fugacity or vapor pressure of pure liquid solute (P_L^S) at the temperature (and strictly the pressure) of the system. This raises a problem for solid solutes such as naphthalene which can not exist as liquids at 25°C. The reference fugacity is then the subcooled liquid vapor pressure, shown in Figure 5.2 earlier, which can only be estimated, not measured. We return to this issue shortly.

The activity coefficient γ_w is defined on a "Raoult's law" basis such that γ_i is 1.0 when x_i is 1.0. In most cases γ_i values exceed 1.0, and for hydrophobic chemicals, values may be in the millions. Again we return to this issue later.

Activity coefficients can be sources of confusion because they can be defined in several ways. The Raoult's law convention is that when x is 1, γ is 1, i.e., pure liquid chemical has unit activity. Often it is not possible to obtain pure liquid chemical (e.g., O_2 or naphthalene at 25°C) so P_L^S can not be measured. The Henry's law convention may then be used that γ is 1 when x is zero and an infinite dilution value of γf_R is defined. The final result is that electrolyte chemists may define γ as being unity at a convenient unit concentration. It is important to qualify γ as to its parentage.

For a solute in water we can thus deduce Z_w as

$$Z_w = 1/v_w \gamma_i P_L^S$$

We thus require a knowledge of γ_i and P_L^S. A short cut method for determining Z_w is to recall that

$$K_{AW} = Z_A/Z_w = H/RT = P^S/C^S RT$$

Therefore $Z_w = Z_A RT/H = 1/H = C^S/P^S$ (since $Z_A = 1/RT$), in which case the C^S and P^S values are for either both liquid, or both solid states.

For liquid solutes such as benzene it is now apparent that

$$H = v_w \gamma_i P_L^S \text{ and } C_L^S = 1/v_w \gamma_i$$

Indeed it is from measured solubilities (C_L^S) that γ_i may be estimated.

For solid solutes such as naphthalene the equations are more complex in that H is P_S^S/C_S^S, a ratio of measurable quantities or P_L^S/C_L^S a ratio of unmeasurable quantities. We define F, the fugacity ratio, as P_S^S/P_L^S or C_S^S/C_L^S, thus

$$H = v_w \gamma_i P_S^S/F \text{ and } C_S^S = F/v_w \gamma_i$$

Calculation of γ from solubility or from H thus requires an estimate of F. This was illustrated earlier in Figure 5.2.

F is 1.0 at the triple point and falls to lower values at lower temperatures. It can be estimated approximately from (Yalkowsky 1979)

$$F = \exp(6.79(1 - T_M/T))$$

where T_M is the melting point temperature and T is the lower system temperature. This contains the assumption that the entropy of fusion at the melting point is 56.5 J/mol. K, i.e., 6.79 R J/mol.K.

In summary, we can now deduce Z_W either as 1/H where H is P_S^S/C_S^S or P_L^S/C_L^S or from $1/v_w\gamma_iP_L^S$ where γ_i can be estimated from the liquid solubility C_L^S, or from the solid solubility C_S^S and F, which in turn can be deduced from the melting point. This is best illustrated by examples.

Worked Example 5.10

Deduce all relevant thermodynamic air-water partitioning properties for benzene (liquid) and naphthalene (a solid of melting point 80°C) at 25°C.

Benzene:

$Z_A = 1/RT = 1/8.314 \times 298 = 4.04 \times 10^{-4}$
vapor pressure $= 12700$ Pa (P_L^S)
molecular mass $= 78$ g/mol
solubility $= 1780$ g/m$^3 = 22.8$ mol/m^3 (C_L^S)
Activity coefficient $\gamma_i = 1/v_wC_L^S = 1/18 \times 10^{-6} \times 22.8 = 2430$
$H = P_L^S/C_L^S = 557$ also $= v_w\gamma_iP_L^S$
$Z_W = 1/H$ or $1/v_w\gamma_iP_L^S = 0.0018$
$K_{AW} = H/RT$ or $Z_A/Z_W = 0.22$

Naphthalene:

$Z_A = 4.04 \times 10^{-4}$ as before
$F = \exp(6.79(1 - 353/298)) = 0.286$
$P_S^S = 10.4$ Pa $P_L^S = P_S^S/F = 36.4$
Solubility $= 31.7$ g/m^3, molecular mass $= 128$ g/mol
$C_S^S = 31.7/128 = 0.25$ mol/m^3 $C_L^S = 0.87$ mol/m^3, i.e. C_S^S/F
$\gamma_i = 1/v_wC_L^S = 64000$

$$H = P_S^S/C_S^S \text{ or } P_L^S/C_L^S = 42$$
$$Z_w = 1/H \text{ or } 0.024 = 1/v_w\gamma_i P_L^S$$
$$K_{AW} = H/RT \text{ or } Z_A/Z_w = 0.017$$

Note that naphthalene has a higher activity coefficient γ_i corresponding to its lower solubility and greater hydrophobicity.

For a solute such as acetone which is miscible with water, no C^S value exists, thus γ must be estimated from a measured H or another data source. Generally such solutes have γ values less than 20 and are fairly "ideal" in solution in water.

Since Z_w is C^S/P^S, substances which have large Z_w values have (as expected) high water solubilities and/or low vapor pressures. Phenols are examples. Substances which have either (or both) low solubility and high vapor pressures tend to partition out of water. Freons partition into air by virtue of their high P^S values (low boiling points), while PCBs and DDT partition into air and organic media by virtue of their low C^S values.

Z_w may be concentration dependent, especially at high concentrations, because of the dependence of γ_i on x, which may be expressed by an activity coefficient equation of the Margules type. These equations usually reduce to a form

$$\ln\gamma_i = \ln\gamma_0.(1-x)^2$$

where γ_0 is the value of the activity coefficient when mole fraction x approaches zero. Only when x is large, i.e., a few percent, does γ_i vary significantly from γ_0; thus, we can usually assume that γ is constant. If variation is suspected, Z can be allowed to vary in a form indicated by the above equation.

Z is temperature dependent because H and its constituent terms v_w, γ_i and especially vapor pressure, are all temperature dependent. The variation of v_w is attributable to the slight thermal expansion of water. The temperature dependence of γ_i can be characterized by the (excess) enthalpy of mixing of solute in water. The dependence of f_R on temperature is characterized by the enthalpy of vaporization of solute from liquid to vapor. The total enthalpy is thus the sum of these terms, or the enthalpy change from solution to vapor. Usually these enthalpies are of different sign, i.e., γ_i usually decreases with increasing T (C^S rises), and P^S always increases with increasing T. The net effect is usually an increase in H with temperature. It is usually necessary to determine H experimentally, either directly, or from solubilities and vapor pressures as was discussed earlier.

In general, for a phase transition such as volatilization or dissolution the temperature dependence is given by an Antoine or Clapeyron type of equation

$$\ln P = A - B/T$$

where P is vapor pressure, solubility, or Henry's law constant. The quantities A and B are chemical-specific constants and T is absolute temperature. It can be shown from the Clapeyron-Clausius relationship that B is $\Delta H/R$ where ΔH is the enthalpy of transition (J/mol) and R is the gas constant (8.314 J/mol.K). It follows that at two temperatures T_1 and T_2

$$\ln (P_1/P_2) = -B(1/T_1 - 1/T_2) = -\Delta H/R(1/T_1 - 1/T_2)$$
$$= \Delta H(T_1 - T_2)/(RT_1T_2)$$

Note that ΔH for an endothermic process is positive; thus, if $T_1 > T_2$, then $P_1 > P_2$. A useful "rule" is that the temperature increase to double P is 0.693 $RT^2/\Delta H$.

Experimentally, P is measured as a function of T and the data fitted to the equation usually using base 10 logs, which alters the value of B by the factor 2.303. Often the equation is modified to the Antoine form, introducing a third constant C

$$\ln P = A - B/(T + C)$$

Z_W may also be pH dependent. A classic case is pentachlorophenol (PCP) which dissociates into the phenate anion and H^+, especially at high pH. The partition and distribution coefficients thus differ. Its acid dissociation constant pKa is 4.8, implying that at pH of 4.8 there is 50% dissociation. At higher pH the basic conditions promote more dissociation and the fraction of PCP present as neutral compound may be quite small. As a result, distribution coefficients from water to air and other liquids such as octanol are pH dependent and Z_W also changes. If only the neutral species partitioned (as probably applies to air), the calculation of distribution coefficients would simply involve estimation of the ratio of neutral N to total T (neutral and ionic A) species from the Henderson-Hasselbach relationship

$$T = N + A$$
$$Ka = H + A/N$$
$$\text{thus } T = N(1+A/N) = N(1+Ka/H +) = N(1+10^{(pH-pKa)})$$

For example, at pH 6.0 the ratio T/N is 17, i.e., 94% is in ionic form and 6% is in neutral form. The air to water distribution coefficient or Z_A/Z_W will probably be 1/17th the partition coefficient value for the neutral species, as would apply at low pH. Ionization thus has the effect of increasing Z_W and reducing evaporation.

Partitioning between water and liquids such as octanol is more complex because both ionic and neutral species may partition.

The text by Lyman et al. (1982) outlines methods of estimating and using pKa values.

5.6.6 Z for Sorbed Phases (e.g., Soil or Sediment)

The simplest method of deducing Z is to relate it to Z for water using the dimensionless partition coefficient. This coefficient can be determined from one of several isotherm equations (Linear, Langmuir, Freundlich, or BET). In most cases the data can be represented, at least at low concentration, by the linear expression

$$C_S = K_{SW}C_W$$

where C_S is the sorbed (concentration mol/m^3 of sorbent), C_W is the water phase concentration (mol/m^3 of water), and K_{SW} is a dimensionless partition coefficient. In the nonlinear equations K_{SW} is a function of concentration. As was discussed earlier, sorbed concentrations are usually expressed as amount of solute (sorbate) per unit mass of sorbent since this avoids measuring, or assuming, a sorbent density. K_P is then defined with units of L/kg and is related to K_{SW} through the sorbent density ρ_S kg/L (or g/cm^3 or Mg/m^3) as

$$K_{SW} = \rho_S K_P$$

It should be noted that the units of solute amount (mg or g or mol) cancel and do not enter K_P, thus any convenient unit can be used.

Since K_{SW} is Z_S/Z_W it follows that

$$Z_S = Z_W K_{SW} = \rho_S K_P/H$$

This definition does not preclude the possibility that K_p is concentration dependent. An allowance for this dependence can be included using an "isotherm" equation. For example, if the Langmuir isotherm is used with an adsorption coefficient B and a maximum sorbate concentration C_M then

$$C_S = C_W C_M B/(1 + C_W B)$$

thus $\qquad K_p = C_S/C_W = C_M B/(1 + C_W B)$

and $\qquad Z_S = \rho_S C_M B/(1 + C_W B)H$

In this case when C_W is small $C_M B$ approaches K_P, but at higher concentrations when sorption sites become saturated, C_S approaches C_M. K_P and Z_S are then reduced.

The alternative Freundlich isotherm is expressed

$$C_S = B C_W 1/n$$

then $\qquad K_p = C_S/C_W = B C_W^{-(1 - 1/n)}$

If n is close to unity the dependence of K_P and Z_S on C_W is slight.

5.6.7   Z for Biotic Phases (e.g., Fish)

Again, it is convenient to calculate Z_B from the biota-water partition coefficient K_{BW}, which is normally expressed like K_p in units such as L water/kg biota. It follows that

$$Z_B = \rho_B K_{BW}/H$$

Normally ρ_B is equal to the density of water and can be ignored numerically. Care must be taken to identify cases in which K_{BW} is defined on a dry weight basis (rather than the wet weight basis assumed here) or on the basis of a concentration in specific tissues such as lipid. In such cases the appropriate density is the mass of dry biota or of tissue per unit volume of wet biota.

For plant material a similar approach is used by defining a corresponding K_{PW} for plant part-water equilibrium. It appears that the lipid content or "octanol-equivalent" content of plant material such as leaves is of the order of 1%.

5.6.8   Z for the Octanol Phase Z_O

As was discussed earlier, K_{OW} is defined as C_O/C_W thus

$$Z_O = Z_W K_{OW} = K_{OW}/H$$

The utility of Z_O is not, of course, directly in environmental calculations, but Z_O is often closely related to Z_B and to Z_S when sorption is dominated by attachment to organic carbon, or dissolution is in lipid phases.

Although Z_O can be calculated simply as shown above, it is useful to explore its fundamental dependence on activity coefficients. Applying the fugacity equation to chemical distributing between octanol and water (subscripts O and W) at equilibrium gives

$$f = x_W \gamma_W f_R = x_O \gamma_O f_R$$

thus
$$x_W \gamma_W = x_O \gamma_O$$
but
$$C_W = x_W/v_W \text{ and } C_O = x_O/v_O$$
thus
$$K_{OW} = C_O/C_W = x_O v_W/x_W v_O = v_W \gamma_W/v_O \gamma_O$$

Substituting for K_{OW} and H in terms of activity coefficients and molar volumes gives

$$Z_O = 1/(v_O \gamma_O P_L^S)$$

which is exactly analogous to the expression for Z_W.

Now v_W and v_O are fixed by the molecular masses and phase densities, and we would expect that the behavior of organic solutes in octanol will be fairly ideal; thus γ_O probably lies in the range 1 to 10. Since K_{OW} is observed to vary by factors of millions, this must be caused by highly variable values of γ_W. It follows that K_{OW} is essentially a descriptor of γ_W.

But solubility in water is also a descriptor of γ_W, for immiscible organic solutes the relationship being as shown earlier

$$C_L^S = 1/v_W\gamma_W \text{ (liquids)} \qquad C_S^S = F/v_W\gamma_W \text{ (solids)}$$

where F is the fugacity ratio. Substituting for γ_W gives a thermodynamic relationship between K_{OW} and C^S.

$$K_{OW} = 1/v_O\gamma_O C_L^S \text{ or } = F/v_O\gamma_O C_S^S$$

A strong inverse relationship is thus expected (and is observed) between K_{OW} and C^S. Several correlation equations have been developed of the form

$$\log K_{OW} = A - B \log C_L^S \text{ (liquids)}$$
$$\log K_{OW} = A - B \log C_S^S + D(T_M - T) \text{ (solids)}$$

Examination of these equations suggests that B should be about 1.0, and A should be log $(1/v_O\gamma_O)$. The fugacity ratio F can be expressed as shown in Section 5.6.5 as

$$\ln F = 6.79 \ (1 - T_M/T)$$
thus
$$\log F = -6.79 \ (T_M - T)/2.303 \ T$$
$$\approx -0.01 \ (T_M - T) \text{ when } T = 298K$$

It follows that D is expected (and found to have) a value of -0.01. For example, Yalkowsky et al. (1982) obtained a correlation of log solubility vs log K_{OW}, i.e., the inverse of the above as

$$\log S_M = -0.944 \log K_{OW} - 0.01 \ MP + 0.323$$

where S_M is solubility (mol/L) and MP is melting point (°C).

The observation that the coefficient is 0.944 and not 1.00 suggests that as molecular weight increases, γ_W increases steadily and significantly, causing C^S to fall, but γ_O is also affected. The regression of log K_{OW} vs log C^S forces expression of all molecular weight variation on to B. Some equations have been published which do not include the D term. They are best ignored if the data set contains solids.

Figure 5.4, which is taken from Miller et al. (1985) gives plots of C_L^S and K_{OW} as a function of molar volume of solute showing the inverse and proportional dependencies. Obviously molar volume is related to log γ_W, i.e., large molecules have large activity coefficients. The net effect is that $C_L^S.K_{OW}$ is fairly constant and a plot of log C_L^S versus log K_{OW} has a slope of nearly -1. Ultimately, it is likely that as more and better data become available, separate expressions will be developed for γ_W and γ_O as a function of molecular

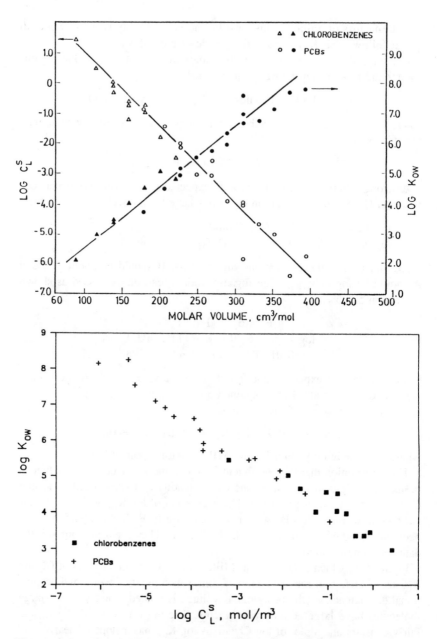

Figure 5.4.   Plots of C_L^S and K_{OW} as a function of molar volume and of C_L^S versus K_{OW} (from Miller et al., 1985).

properties, the values corrected for the presence of water in octanol and octanol in water, and C_L^S and K_{OW} calculated from these corrected activity coefficients.

The reader is cautioned that the literature contains references to calculated log K_{OW} values exceeding 8. These are highly suspect, because it is very difficult to make such measurements.

The question now arises, since K_{OW} and C^S are both merely descriptors of γ_W, why use both? The answer is probably twofold. First, K_{OW} expresses γ_W directly, whereas for solid chemicals C_S^S expresses F/γ_W. There is often doubt about F, especially for high melting point solids. The solid solubility thus depends on γ_W and melting point, which is in turn a reflection of crystal stability. Since crystals of contaminants rarely exist in the environment, the solid solubility and F are only of academic interest. Use of K_{OW} is a simple method of expressing γ_W or hydrophobicity directly.

Second, as was discussed earlier, high molecular weight appears to cause an increase in γ_O. It may also cause a corresponding increase in γ in lipids and in organic matter; thus, K_{OW} may accidentally "correct" for this effect, giving an improved predictive capability.

Ultimately it is likely that data of sufficient quality and quantity will be available so that γ_W and γ_O can be deduced and correlated accurately, and the use of K_{OW} will be unnecessary. Solubility will always be needed as a contributor to H, but its more general use is still impeded by lack of data for F. There is a need to measure and interpret data for K_{OW}, C^S, H, F, and the enthalpies of the phase transitions to build up a sound thermodynamic database for the behavior of environmental contaminants in aqueous and organic solutions.

5.6.9 Z for Aerosols

The Z value for aerosols Z_X can be deduced from the aerosol/air partition coefficient K_{XA}, which in turn is related to subcooled liquid vapor pressure, (not the solid vapor pressure) i.e.,

$$K_{XA} = Z_X/Z_A$$

It follows that Z_X is given by

$$Z_X = Z_A K_{XA} = 6 \times 10^6/P_L^S RT$$

Since P_L^S is highly temperature dependent, Z_X, and thus the fraction of chemical sorbed to aerosol particles, are also very temperature dependent.

Alternatively one of the A.TSP/F correlations can be used to deduce K_{XA} and hence Z_X.

5.6.10   Z for Pure Solute Phases

The fugacity of a pure solute is usually its vapor pressure P^S, and its "concentration" is the reciprocal of its molar volume v_S (m³/mol) (typically 10^{-4}m³/mol) thus

$$C = (1/v_S) = Z_P f = Z_P P^S$$
thus
$$Z_P = 1/P^S v_S$$

Although it may appear environmentally irrelevant to introduce Z_P, there are situations in which it is used. If there is a spill of PCB or an oil into water of sufficient quantity that the solubility is exceeded, at least locally, the environmental partitioning calculations may involve the use of volumes and Z values for water, air, sediment, biota, and a separate pure solute phase. Indeed, early in the spill history most of the solute will be present in this phase. The difference in behavior of this and other phases is that the pure phase fugacity (and of course concentration) remains constant, and as the chemical migrates out of the pure phase, the phase *volume* decreases until it becomes zero at total dissolution or evaporation. In the case of other phases, the *concentration* changes at approximately constant volume as a result of migration.

Interestingly, a pure solute/aerosol partition coefficient can be deduced as Z_P/Z_X or $(1/P^S v_S)/(6 \times 10^6/P^S RT)$. The vapor pressure cancels, v_S is typically 10^{-4} m³/mol, and RT is typically 2500 Pa.m³/mol; thus, K_{PX} is of the order of 4, i.e., aerosols behave like absorbent sponges achieving concentrations of 25% pure solute!

Similarly, Z_P/Z_O (pure solute-octanol) can be shown to be

$$Z_P/Z_O = 1/(P_L^S v_S)/(1/v_O \gamma_O P_L^S) = v_O \gamma_O/v_S$$

Again, if the solute is organic, γ_O is expected to lie in the range 1 to 10; thus, Z_P/Z_O will also lie in this range. To a first approximation the capacities of pure solute, octanol and aerosol particles are similar.

Figure 5.5 illustrates and summarizes the relationships between these Z values and partition coefficients.

5.6.11   Z for Other Phases

It is conceivable that a need may arise to define other Z values. In principle the approach is to use either the fundamental fugacity equation or use a dimensionless partition coefficient to deduce Z from another phase of known Z.

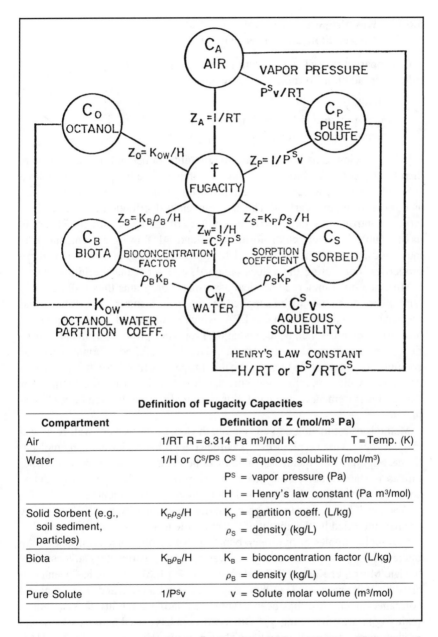

Figure 5.5. Relationships between Z values and partition coefficients and summary of Z value definitions.

5.6.12 Key Properties Controlling Z Values, Correlation, and Measurement

It is apparent that when calculating Z values the key properties of a chemical are, in addition to molecular structure and mass,

> solubility in water,
> vapor pressure (i.e., solubility in air),
> octanol-water partition coefficient.

If K_{OW} is viewed as being a ratio of solubility in octanol to that in water, then $K_{OW}C_L^S$ is the solubility in octanol. We thus reduce the properties to three solubilities, in air, water, and octanol as a surrogate for organic phases such as lipids and organic carbon. The environmental distribution of chemicals is largely controlled by the relative magnitudes of these solubilities. Phenol is most soluble in water, chloroform in air, and DDT in organic media. It is thus critically important to have available reliable methods of determining these properties, correlating the values obtained, and accessing the information.

Most new-to-commerce chemical assessment programs thus require these (and other) properties to be supplied by the manufacturer. The notable and regrettable exception is the U.S. EPA, which demands only molecular structure. EPA scientists then deduce what the properties probably are, and if justified, they can demand measurements. This has resulted in enthusiasm for computer methods of deducing probable properties from structure, which is certainly a valid scientific endeavor, and it is fun, but it is no substitute for careful measurement. Calculated values should always be treated with suspicion.

Methods of estimating and correlating the key properties are well reviewed in the text by Lyman et al. (1982). Dearden (1990) has reviewed many molecular descriptors, i.e., quantities which can be deduced from molecular structure as a means of correlating structure and properties. It is helpful to review some of the more useful ideas and methods of measurement.

Water solubility appears to be primarily controlled by molecular volume or area, modified by the ability of the molecule to interact with the water by, for example, dipoles or hydrogen bonding. Numerous correlations exist. An interesting approach is the use of "solvatochromic" parameters advocated by the late Mortimer Kamlet (Kamlet et al., 1988, 1986), but this has been criticized by Yalkowsky et al. (1988). Since K_{OW} can be estimated from fragment constants, much effort has been devoted to relating solubility to K_{OW}, but in reality both are correlating γ_W. Other workers have used activity coefficient prediction schemes such as regular solution theory, UNIFAC, and UNIQUAC to correlate and predict solubility.

It must be noted that solubility is temperature sensitive; it is affected by the presence of electrolytes (the salting-out effect) and organic co-solvents. Apparent solubility is affected by the presence of sorbing phases, humic acids, and surfactants.

The preferred method of measuring solubility for very soluble chemicals (i.e., >100 mg/L), is the simple shake-flask or two phase stirring method. For less soluble chemicals the "generator column" method is preferred in which water is slowly flowed through a column containing the chemical "plated" on a solid support such as glass beads or firebrick. A slow stirring method can also be used, but extreme care must be taken to avoid forming colloidal solute particles since these can easily establish a stable concentration of 10 mg/L, thus totally obscuring an actual solubility of, say, 0.1 mg/L.

Vapor pressure is apparently controlled by the size of the molecule and especially the energy needed to separate a molecule from its neighbors and isolate it in the vapor phase. As was discussed earlier, vapor pressures are very temperature sensitive. Several correlation schemes exist, many employing boiling point (which is the temperature at which the vapor pressure is 101 kPa) and critical temperature. Burkhard et al. (1985) have reviewed 11 methods of predicting vapor pressures, one of the simpler of which was devised by Mackay et al. (1982). In general, the more data that are available for a chemical, the better the prediction. Ideally, data should be reported or estimated over a wide range of temperature, i.e., -25 to $+50°C$.

The preferred measurement technique is the "gas saturation" method in which gas (usually N_2) is flowed through a column containing the chemical to achieve saturation, e.g., Spencer and Cliath (1969), Rordorf (1985, 1987), Hansen and Eckert (1986), Sonnefeld and Zoller (1983), and Westcott et al. (1981). If vapor pressure data are available for structurally similar chemicals, a convenient method is to estimate vapor pressure from GC retention time (e.g., Foreman and Bidleman, 1985). A very simple method is to observe relative evaporation rates of pure chemical (Dobbs and Cull, 1982).

5.7 MAXIMUM FUGACITY AND SATURATION

In partitioning calculations when fugacities are calculated routinely, it is advisable to check that the value deduced is lower than the fugacity of the pure phase, i.e., the solid or liquid fugacity or in the case of gases, of atmospheric pressure. If these fugacities are exceeded, supersaturation has occurred, a "maximum permissible fugacity" has been exceeded, and the system will automatically "precipitate" a pure solute phase until the fugacity drops to the saturation value. It is possible to calculate inadvertently and use (i.e., misuse) these "over-maximum" fugacities. It is also noteworthy that the maximum Z value that a noninteracting solute can ever achieve is that of the pure phase Z_p. It may be useful to calculate Z_p to ensure that no mistakes have been made by grossly overestimating other Z values.

5.8   CHEMICALS OF NEGLIGIBLE VOLATILITY

A problem arises when calculating values of the fugacity or fugacity capacity of solutes which have a negligible or zero vapor pressure. Thermodynamically, the problem is that of determining the reference fugacity. The practical problem may be that no values of vapor pressure or air-water partition coefficients are published, or even exist. Examples are metallic ions, inorganic materials such as calcium carbonate or silica, and polymeric or high molecular weight substances including carbohydrates and proteins. Intuitively, no vapor pressure determination is needed (or may be possible) because the substance does not partition into the atmosphere, i.e., its "solubility" in air is effectively zero. Ironically, its air fugacity capacity can be calculated as (1/RT), but all the other (and the only useful) Z values cannot be calculated since H cannot be determined, and indeed may be zero. Apparently all the non-air Z values are infinite, or at least are indeterminably large.

This difficulty is more apparent than real, and is a consequence of the selection of fugacity rather than activity as an equilibrium criterion. There are two remedies.

First, and convenient, but somewhat dishonest, is to assume a fictitious, reasonable, but small, value for vapor pressure (such as 10^{-6} Pa), and proceed through the calculations using this value. The result will be that Z for air will be very small compared to the other phases, and negligible concentrations will result in the air. It is obviously essential to recognize that these air concentrations are fictitious and erroneous. The relative values of the other concentrations and Z values will be correct, but the absolute fugacities will be meaningless.

Second, less convenient, but more honest, is to select a new equilibrium criterion. We can illustrate this for air, water, and another phase(s) by equating fugacities as follows.

$$f = C_A/Z_A = C_W/Z_W = C_S/Z_S$$
$$f = C_A RT = C_W P^S/C^S = C_S P^S/C^S K_{SW}$$

We can divide through by P^S to give

$$f/P^S = a = C_A RT/P^S = C_W/C^S = C_S/C^S K_{SW}$$

The equilibrium criterion is now a, an activity which is dimensionless and is the ratio of fugacity to vapor pressure. The new Z values with units of mol/m^3 can be defined as

$$\text{air } Z = P^S/RT \quad \text{water } Z = C^S \quad \text{sorbed } Z = C^S K_{SW}$$

A saturated solution thus has an activity of 1.0. A zero or near zero vapor pressure can be used to calculate Z for air as zero or near zero.

In some cases we may have to go further because we are also uncertain about C^S. The simple expedient is then to multiply through by C^S to give yet another equilibrium criterion A as

$$fC^S/P^S = A = C_A RTC^S/P^S = C_W = C_S/K_{SW}$$

or for air $Z = P^S/RTC^S$, water $Z = 1.0$, sorbed phase $Z = K_{SW}$.

We are now using the water concentration or the equivalent equilibrium water concentration as the criterion of equilibrium. This has been termed the "aquivalent concentration" (Mackay and Diamond 1989) and can be used for metals in ionic form, for which solubility is meaningless. The concentration C is now A.Z.

5.9 SOME ENVIRONMENTAL IMPLICATIONS

Viewing the behavior of a solute in the environment in terms of Z introduces new and valuable insights. A solute tends to migrate into (or stay in) the phase of largest Z. Thus SO_2 and phenol tend to migrate into water, freons into air, and DDT into sediment or biota. The phenomenon of bioconcentration is merely a manifestation of a Z value in biota which is much higher (by the bioconcentration factor) than Z in the water. Occasionally, a solute such as inorganic mercury changes its chemical form, becoming organometallic (e.g., methylmercury). Its Z values change and the mercury now sets out on a new environmental journey with a destination of the new phase in which Z is now large. In the case of mercury, the ionic form will sorb to sediments or dissolve in water, but will not appreciably bioconcentrate. The organic form experiences a large Z in biota and will bioconcentrate.

Some solutes such as DDT or PCBs have very low Z values in water because of their highly hydrophobic nature, i.e., they exert a high fugacity even at low concentration, reflecting a large "escaping tendency." They will therefore migrate readily into any neighboring phase such as sediment, biota, or the atmosphere. Atmospheric transport should thus be no surprise and the contamination of biota in areas remote from sites of use is expected. With this hindsight, it is not surprising that these substances are found in the tissues of Arctic bears and Antarctic penguins!

From the environmental monitoring and analysis viewpoint, it is preferable to sample and analyze phases in which Z is large because it is in these phases that concentrations should be greatest and thus easier to determine accurately. When monitoring for PCBs in lakes it is thus common to sample sediments or fish rather than water, since the expected water concentrations are very low. Likewise, those concerned with PCB behavior in the atmosphere may concentrate on measuring the PCBs on aerosols or in rainfall containing aerosols, since concentrations are higher than in the air.

In general, when assessing the likely environmental behavior of a new chemical it is useful to calculate the various Z values and from them identify the larger ones, since it is likely that the high Z compartments are the most important. It is no coincidence that solutes such as halogenated hydrocarbons, about which there is greatest public concern, have high Z values in humans!

It should be borne in mind that when calculating the environmental behavior of a solute, Z values are needed only for the phases of concern. For example, if no atmospheric partitioning is considered, it is not necessary to know the air-water partition coefficient or H. An arbitrary value of H of 1.0 can be used to define Z for water and other phases. It transpires that H cancels. Intuitively, it is obvious that H, or vapor pressure, plays no role in influencing water-fish-sediment equilibria.

5.10   LEVEL I CALCULATIONS

Recapitulating Example 5.9, it is apparent that we can undertake simple multimedia equilibrium calculations using the equations.

$$M = \Sigma C_i V_i = f \Sigma V_i Z_i$$

thus
$$f = M / \Sigma V_i Z_i \qquad C_i = f Z_i$$
$$m_i = C_i V_i = Z_i V_i f$$

from which
$$M = \Sigma m_i$$

Here M is total moles, m_i is moles in compartment i, and f is the common fugacity which prevails over all compartments. The key properties determining Z_i are molecular mass, solubility in water, K_{OW}, and vapor pressure, as well as the environmental properties of temperature, densities, organic carbon, and lipid contents.

Alternatively, one concentration could be given from which f can be calculated, enabling other equilibrium concentrations to be deduced.

If all concentrations are given, they can be converted to fugacities and compared to determine if equilibrium exists, or to reveal the equilibrium status of the reported values.

The most common calculation is the first or "Level I" type. It is particularly useful for assessing the likely general fate of a chemical in an evaluative environment of the type designed in Chapter 4.

If a mean Z value is to be calculated for a phase containing other phases (e.g., biota in water), the individual Z values are combined in proportion of the volume fractions v,

$$\text{i.e., } Z \text{ mean} = v_1 Z_1 + v_2 Z_2 + \dots$$

Calculation of the equilibrium distribution of a chemical is simple, but it can be tedious. It is very suitable for implementation on a form (like an income tax form), or on a computer. The obvious steps are:

1. Definition of the environment.
2. Input of physical chemical properties.
3. Calculation of Z values.
4. Input of chemical amount.
5. Calculation of fugacity, and hence concentrations and amounts.

To assist these calculations, two "fugacity forms" have been devised. An example of a hypothetical chemical is shown later in Example 5.12, and blank forms are also provided in the appendix for reproduction. These calculations can become quite tedious; it being obvious that they are ideal for implementation on a computer program or spreadsheet. Level I fugacity programs in BASIC language are also provided on the diskette. The output is self explanatory. Users are encouraged to work through examples by hand and then by computer, and vary the media volumes and properties. It is suggested that the partitioning of chemicals in Table 3.2 be examined using this approach, and the results examined. Sensitivity analyses of the results to the assumed properties can be undertaken.

Worked Example 5.11

Calculate, by hand, the distribution of 100 mol of a hypothetical chemical "hypothene" in the simple four compartment unit world in Table 4.2 using the properties below which apply at 27.5°C. Assume organic carbon contents of 2% and 4% in soil and sediment.

(a) Calculation of Z values
Molecular mass 200 g/mol, temperature 300.6 K, solubility 20 g/m³, vapor pressure 1 Pa, log K_{OW} 4.

$$K_{OC} = 0.41 \ K_{OW}$$
$$= 4100$$

Air Z_A = $1/RT$ = $1/8.314 \times 300.6$ = 4×10^{-4}
Water Z_W = $1/H$ = C^S/P^S = $(20/200)/1$ = 0.1
Soil (E) Z_E = $K_{PE}Z_W$ = $Z_W y_E K_{OC} \rho_E /1000$ = $0.1 \times 0.02 \times$
        $4100 \times 1500/1000$ = 12.3
Sediment Z_S = $K_{PS}Z_W$ = $Z_W y_S K_{OC} \rho_S /1000$ = $0.1 \times 0.04 \times$
        $4100 \times 1500/1000$ = 24.6

(b) Calculation of fugacity f

$$\Sigma VZ = V_A Z_A + V_W Z_W + V_E Z_E + V_S Z_S$$
$$= 6 \times 10^9 \times 4 \times 10^{-4} + 7 \times 10^6 \times 0.1 + 45000 \times 12.3 +$$
$$21000 \times 24.6 = 4170000$$

f   = $M/\Sigma VZ$ = $100/4170000$ = 2.4×10^{-5} Pa

(c) Calculation of concentrations and amounts

C_A	= fZ_A = 9.6×10^{-9}	M_A	= $C_A V_A$	= 57.5
C_W	= fZ_W = 2.4×10^{-6}	M_W	= $C_W V_W$	= 16.8
C_E	= fZ_E = 2.9×10^{-4}	M_E	= $C_E V_E$	= 13.3
C_S	= fZ_S = 5.9×10^{-4}	M_S	= $C_S V_S$	= 12.4
				100.0

The corresponding computer program output for this example is given in Figure 5.6.

Worked Example 5.12

Recalculate, using the fugacity forms, Example 5.11, but include suspended sediment and biota (fish) at the recommended volume fractions, and increase the hydrophobicity of the chemical to a log K_{OW} of 5, thus driving more chemical into organic carbon and lipid phases.

The results are given in Figures 5.7 and 5.8.

FUGACITY LEVEL I CALCULATION

Properties of HYPOTHENE

Temperature deg C	27.5
Molecular mass g/mol	200
Vapor pressure Pa	1
Solubility g/m³	20
Solubility mol/m³	.1
Henry's law constant Pa.m³/mol	10
Log octanol-water p-coefficient	4
Octanol-water partn-coefficient	10000
Organic C-water ptn-coefficient	4100
Air-water partition coefficient	4.00 E-03
Soil-water partition coefficient	123
Sedt-water partition coefficient	246
Amount of chemical moles	100
Fugacity Pa	2.397 E-05
Total of VZ products	4170480

Phase properties and compositions

Phase	Air	Water	Soil	Sediment
Volume m³	6E+09	7000000	45000	21000
Density kg/m³	1.175	1000	1500	1500
Frn org carb	0	0	.02	.04
Z mol/m³.Pa	4.000 E-04	.1	12.3	24.6
VZ mol/Pa	2400380	700000	553500	516600
Fugacity Pa	2.397 E-05	2.397 E-05	2.397 E-05	2.397 E-05
Conc mol/m³	9.592 E-09	2.397 E-06	2.949 E-04	5.898 E-04
Conc g/m³	1.918 E-06	4.795 E-04	5.898 E-02	.117
Conc µg/g	1.632 E-03	4.795 E-04	3.932 E-02	7.864 E-02
Amount mol	57.556	16.784	13.271	12.387
Amount %	57.556	16.784	13.271	12.387

Figure 5.6. **Specimen output of a Level I calculation corresponding to Worked Example 5.11.**

FUGACITY FORM 1			Z VALUES

CHEMICAL: $HYPOTHENE-2$

Temperature: 27.5 .c 300.6 K   RT = 8.314 x 3006 = 2500

Molecular mass: 200 g/mol (w)

Water solubility: 20 g/m³ or mg/L 0.1 mol/m³ (C^S)

Vapor pressure: 1.0 Pa (P^S) 7.5×10^{-3} mm Hg 9.9×10^{-6} atm

Log K$_{OW}$: 5 K$_{OW}$ $100,000$

AIR-WATER PARTITION COEFFICIENTS AND Z VALUES

H = P^S/C^S = 10 Pa.m³/mol

K_{AW} = H/RT = 0.004

Z_A (air) = 1/RT = 4.0×10^{-4} Air density = 0.029 x 101325/RT

Z_W (water) = 1/H = 0.1 = 1.17 kg/m³

OTHER PHASES

Name	SOIL	SEDIMENT	SUSPENDED SEDIMENT	FISH
Density ρ kg/m³	1500	1500	1500	1000
Organic carbon or lipid content φ g/g	0.02	0.04	0.04	0.048
K$_{OC}$ or K$_{OL}$ (1)	41000	41000	41000	100000
K$_P$ = φK$_{OC}$ or φK$_{OL}$	820	1640	1640	4800
K$_{PW}$ = K$_P$ρ/1000	1230	2460	2460	4800
Z$_P$ = K$_{PW}$·Z$_W$	123	246	246	480

(1) e.g. K$_{OC}$ = 0.41 K$_{OW}$, K$_{OL}$ = K$_{OW}$

Figure 5.7. Fugacity Form 1 illustrating the calculation of Z values for a hypothetical chemical for use in a Level I calculation.

FUGACITY FORM 2						LEVEL I
CHEMICAL: $HYPOTHENE-2$ Amount M 100 mol 20 kg						
Compartment	AIR	WATER	SOIL	SEDIMENT	SUSP. SEDT.	FISH	TOTAL
Volume V m³	6×10^9	7×10^6	4.5×10^4	2.1×10^4	35	7	—
Z mol/m³ Pa	4×10^{-4}	0.1	123	246	246	480	—
VZ mol/Pa	2.4×10^6	7×10^5	5.54×10^6	5.17×10^6	8610	3360	1.381×10^7

$$\frac{\text{Amount M mol}}{\Sigma VZ} = \frac{100}{1.381\times10^7} = 7.24\times10^{-6} = \text{FUGACITY f}$$

	AIR	WATER	SOIL	SEDIMENT	SUSP. SEDT.	FISH	TOTAL
C = Zf mol/m³	2.9×10^{-9}	7.2×10^{-7}	8.9×10^{-4}	1.8×10^{-3}	1.8×10^{-3}	3.5×10^{-3}	—
m = CV mol	17.4	5.1	40.0	37.4	0.062	0.024	100
percent	17.4	5.1	40.0	37.4	0.062	0.024	100
C_G g/m³ [1]	5.8×10^{-7}	1.4×10^{-4}	0.178	0.36	0.36	0.7	—
Density ρ kg/m³	1.17	1000	1500	1500	1500	1000	—
C_U μg/g [2]	4.9×10^{-4}	1.4×10^{-4}	0.12	0.24	0.24	0.7	—

[1] C_G = C x Molecular Mass (g/mol)
[2] C_U = C_G x 1000/Density (kg/m³)

Figure 5.8.   Fugacity Form 2 illustrating a Level I calculation and showing the environmental distribution of the hypothetical chemical in Figure 5.7.

5.11   SUMMARY

In this chapter we have introduced the concept of equilibrium existing between phases and have shown that this concept is essentially dictated by the laws of thermodynamics. Fortunately, we do not need to use, or even understand, the thermodynamic equations on which equilibrium relationships are based. However, it is useful to use these relationships for purposes such as correlation of partition coefficients. It transpires that there are two approaches which can be used to conduct equilibrium calculations. First, is to develop and use empirical correlations for partition coefficients. Using these coefficients it is possible to calculate the partitioning of the chemical in a multimedia environment.

The second approach, which we prefer, is to use an equilibrium criterion such as fugacity or activity, or in the case of involatile chemicals, an aquivalent concentration. The criterion can be related to concentration for each chemical, and for each medium, using a proportionality constant or Z value. The Z value can be calculated from fundamental equations or from partition coefficients. We have established recipes for the various Z values in these media using information about the nature of the media and the physical chemical properties of the substance of interest. This enables us to undertake simple multimedia or Level I partitioning calculations.

6   ENVIRONMENTAL LOSS MECHANISMS

6.1   INTRODUCTION

In Level I calculations, it is assumed that the chemical is conserved, i.e., it is neither destroyed by reactions, nor conveyed out of the evaluative environment by flows in phases such as air and water. These assumptions can be quite misleading when assessing the impact of a given discharge or emission of chemical.

First, if a chemical, such as phenol, is reactive and survives for only 10 hours as a result of its susceptibility to rapid biodegradation, it must pose less of a threat than PCBs, which may survive for 10 years. But the Level I calculation treats them identically in terms of persistence.

Second, some chemical may leave the area of discharge rapidly as a result of evaporation into air and flow, or advection in winds. The contamination problem is not necessarily solved, it is merely relocated. It is important to know if this will occur.

Third, it is possible that in a given region, local contamination is largely a result of inflow of chemical from upwind or upstream regions. Local efforts to reduce contamination by controlling local sources may thus be frustrated because most of the chemical is inadvertently imported. This problem is at the heart of the Canada-U.S., and Scandinavia-Germany-U.K. squabbles over acid rain.

It must be emphasized that a chemical degraded is not necessarily a problem solved. Toxicologists rarely miss an opportunity to point out that degrading reactions can give rise to new, troublesome chemicals. They point to reactions, such as mercury methylation, or benzo(a)pyrene oxidation, in which the product of the reaction is much more harmful than the parent compound. However, we will be content here to treat only the parent compound. Assessment of degradation products is best done separately by having one chemical's degradation product treated as another's formation. In this chapter, we address these issues and devise methods of calculating the effect of advective inflow and outflow, and degrading reactions on local chemical fate and exposures.

A key concept in this discussion which was introduced earlier is variously termed "persistence" or "lifetime" or "residence time" or "detention time" of the chemical.

In a steady-state system, as shown in Figure 6.1, if chemical is introduced at a rate of E mol/h, then the rate of removal must also be E mol/h; otherwise net accumulation or depletion will occur. If the amount in the system is M mol, then on the average, the amount of time each molecule spends in the system will be M/E hours. This time, t, is a residence time or detention time or persistence. Clearly, if chemical persists for longer, there will be more of it in the system. The key equation is

$$t = M/E \text{ or } M = tE$$

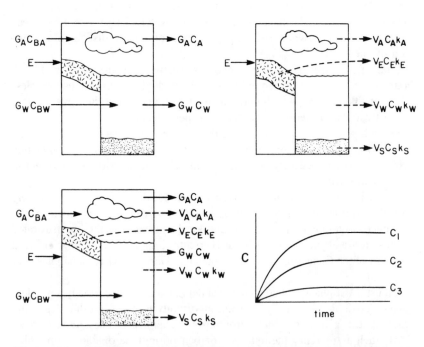

Figure 6.1.   Diagram of a steady state evaluative environment subject to (a) advective flow, (b) degrading reactions, (c) both, and (d) the time course to steady state.

This idea is most commonly applied to retention time in lakes. A lake may have a volume of 100,000 m³. If it receives an inflow of 1000 m³ per day, then the retention time is 100,000/1,000 or 100 days. A mean retention time of 100 days does not imply that all water will spend 100 days in the lake. Some will bypass in only 10 days, and some will persist in backwaters for 1,000 days, but on the average, the residence time will be 100 days.

The reason that this concept is so important is that chemicals experience highly variable lifetimes, ranging from hours to decades. As a result, the amount of chemical present in the environment, i.e., the inventory of chemical, varies greatly. We tend to be most concerned about persistent and toxic chemicals (large t) because even relatively small emission rates (E) can result in large amounts (M) present in the environment. This translates into high concentrations and possibly significant adverse effects. A further consideration is that chemicals which survive for prolonged periods in the environment have the opportunity to make long, and often tortuous, journeys. If applied to soil, they may evaporate, migrate onto atmospheric particles, deposit on vegetation, be eaten by cows, transferred to milk, and consumed by humans. They may migrate up the food chain from water to plankton to fish to gulls or otters. Short-lived chemicals rarely survive long enough to undertake such adventures (or misadventures).

This lengthy justification leads to the conclusion that, if we are going to discharge a chemical into the environment, it is prudent to know

(a) how long the chemical will survive (t) and
(b) what causes its removal or "death."

This latter knowledge is useful because situations are likely in which a customary removal mechanism does not apply. For example, a chemical may be subject to rapid photolysis, but this is not of much relevance in long dark arctic winters, or in deep murky sediments.

In the process of quantifying this effect, we will introduce rate constants, advective flow rates, and ultimately, using the fugacity concept, quantities called D-values, which prove to be immensely convenient. Indeed, armed with Z-values and D-values, the environmental scientist has a powerful set of calculation and interpretive tools.

It transpires that there are two primary removal mechanisms, advection and reaction, which we first discuss individually, then in combination.

6.2 ADVECTION

Strangely, "advection" is a word absent from many dictionaries, so a definition is appropriate. It means the movement of chemical by virtue of its presence in a medium which happens to be flowing. A lazy canoeist is thus advected down a river. PCBs are advected from Chicago to Buffalo in a westerly wind. SO_2 may be advected from U.K. power plants to Sweden. The rate of advection of a chemical N (mol/h) is simply the product of the flowrate of the advecting medium, G (m^3/h) and the concentration of chemical in that medium, C (mol/m^3)

$$i.e., N = GC$$

Thus, if there is river flow of 1,000 m^3/h (G) from A to B of water containing 0.3 mol/m^3 (C) of chemical, then the corresponding flow of chemical is 300 mol/h (N).

Turning to the evaluative environment, it is apparent that the primary candidate advective phases are air and water. If, for example, there was air flow into the evaluative environment of 10^8 m^3/h, then since the volume of the air in the evaluative environment is 6×10^9 m^3, the residence time will be 60 hours, or 2.5 days. Likewise, the flow of 100 m^3/h of water into 70,000 m^3 of water results in a residence time of 700 hours, or 29 days. It is usually easier to remember the residence time than the flow rate.

Level II Algebra, Partition Coefficients

If we decree that our evaluative environment is at steady state, then inflows must equal outflows, and the advective inflow rates, designated G m^3/h, must also be outflow rates. If the concentrations of chemical in the phase of the evaluative environment is C mol/m^3, then the outflow rate will be GC mol/h. This concept is often termed the continuously stirred tank reactor, or CSTR, assumption. The idea is that, if a volume of phase, for example air, is well stirred, then if some of that phase is removed, it must have a concentration equal to that of the phase as a whole. If chemical is introduced to the phase at a different concentration, it experiences an immediate change in concentration to that of the well mixed, or CSTR, value. The concentration experienced by the chemical then remains constant until the chemical is removed. The key point is that the outflow concentration equals the prevailing concentration. This concept greatly simplifies the algebra of steady state systems. Essentially, we treat air, water, and other phases as being well mixed CSTRs in which the outflow concentration equals the prevailing concentration. We can now consider an evaluative environment in which there is inflow and outflow of chemical in air and water. It is convenient at this stage to ignore water particles, fish, and aerosols, and assume that the material flowing into the evaluative environment is pure air and pure water contaminated only by chemical. Since steady state applies, as shown in the first diagram of Figure 6.1, the inflow and outflow rates are equal and a mass balance can be assembled. The total influx of chemical is at a rate $G_A C_{BA}$ in air, and $G_W C_{BW}$ in water, these concentrations being the "background" values. There may also be emissions into the evaluative environment at a rate E. The total influx is thus

$$I = E + G_A C_{BA} + G_W C_{BW} \text{ mol/h}$$

Now, the concentrations within the environment adjust to values C_A and C_W in air and water. Thus, the outflow rates must be $G_A C_A$ and $G_W C_W$. These environmental and outflow concentrations could be constrained by equilibrium considerations; for example, they may be related through partition coefficients or through Z-values to a common fugacity.

This enables us to conceive of, and define, our first Level II calculation, in which we assume equilibrium and steady state to apply, inputs by emission and advection are balanced exactly by advective emissions, and equilibrium exists throughout the evaluative environment. All the phases are behaving like individual CSTRs.

Of course, starting with a clean environment and introducing these inflows, it would take the system some time to reach steady state conditions, as shown in Figure 6.1. But we are not concerned, at this stage, with how long it takes to reach steady state; only the conditions which apply at steady state. We can thus develop the following equations, using partition coefficients or fugacities for a simple two compartment system.

$$I = E + G_A C_{BA} + G_W C_{BW} = G_A C_A + G_W C_W$$

Using partition coefficients

$$C_A = K_{AW} C_W$$

therefore $I = C_W[G_A K_{AW} + G_W]$ and $C_W = I/[G_A K_{AW} + G_W]$

from which other concentrations, amounts (m), and the total amount M can be deduced. The extension to multiple compartment systems is obvious.

Level II Algebra, Fugacity Format

In fugacity format we assume a common fugacity f to apply, thus

$$I = G_A Z_A f + G_W Z_W f = f(G_A Z_A + G_W Z_W)$$
$$f = I/(G_A Z_A + G_W Z_W)$$

or in general $\quad f = I/\Sigma G_i Z_i$

from which the fugacities and all concentrations and amounts can be deduced.

Worked Example 6.1

An evaluative environment consists of 10^4 m^3 air, 100 m^3 water, and 1 m^3 sediment. There is air inflow of 1000 m^3/h and water inflow of 1 m^3/h at respective chemical concentrations of 0.01 mol/m^3 and 1 mol/m^3. The Z values are air 4×10^{-4}; water 0.1; and sediment 1.0. There are also emissions of 4 mol/h. Calculate the fugacity, concentrations, persistence, amounts, and outflow rates.

$$I = E + G_A C_{BA} + G_W C_{BW} = 4 + 1000 \times 0.01 + 1 \times 1 = 15$$
$$\Sigma GZ = 1000 \times 4 \times 10^{-4} + 1 \times 10^{-1} = 0.5, \ f = I/\Sigma GZ = 30$$
$$C_A = 0.012, \ C_W = 3, \ C_S = 30, \ m_A = 120, \ m_W = 300, \ m_S = 30,$$
$$M = 450$$
$$G_A C_A = 12 \ G_W C_W = 3 \ G_S C_S = 0 \ \text{Total} = 15 = I$$
$$t = 450/15 = 30h$$

In this example the total amount of material in the system, M, is 450 mols. But the inflow rate is 15 mols/h; thus the residence time or the persistence of the chemical is clearly 30 hours. This proves to be a very useful time. Note that the air residence time is 10 hours and water residence time is 100 hours. The overall residence time of the chemical is a weighted average, influenced by the extent to which the chemical partitions into the various phases. The sediment has no effect on the fugacity or the outflow rates, but it acts as an inert "reservoir" to influence the total amount present M, and thus the persistence.

D Values

The group GZ, and other groups like it, appear so frequently in later calculations that it is convenient to designate them as D-values,

$$\text{i.e., GZ} = D$$

and the rate, N mol/h, then equals Df. These D values are transport parameters, with units of mol/Pa.h. When multiplied by a fugacity, they give rates of transport. They are thus similar in principle to rate constants, which, when multiplied by concentration C, give a rate. Fast processes have large D-values. We can rewrite the fugacity equation for the evaluative environment in more compact form below as follows.

$$f = I/(D_{AA} + D_{AW}) = I/\Sigma D_{Ai}$$

where D_{AA} is $G_A Z_A$, D_{AW} is $G_W Z_W$, and the first subscript A refers to advection.

In Example 6.1

$$D_{AA} = 0.4 \text{ and } D_{AW} = 0.1 \text{ and } \Sigma D_{Ai} = 0.5$$

thus $\qquad\qquad\qquad f = 15/0.5 = 30$

It is apparent that the air D-value is larger and most significant. D-values can be added when they are multiplied by a common fugacity. Thus, it becomes obvious which D-value, and hence which process, is most important. We can arrive at the same conclusion using partition coefficients, but the algebra is less elegant.

Note that the source of the chemical to the environment (i.e., emission or advection to air or to water) is unimportant, all sources being lumped in I, the overall input. This is because the model formulation decrees that when the chemical enters the environment it immediately achieves an equilibrium distribution and forgets its origin.

Advective Processes

In an evaluative environment, there are several advective flows which potentially convey chemical to and from the environment, namely:

(1) outflow and inflow of air
(2) inflow and outflow of water
(3) inflow and outflow of aerosol particles present in air
(4) inflow and outflow of particles and biota present in water
(5) transport of air from the troposphere to the stratosphere, i.e., vertical movement of air out of the environment
(6) sediment burial, i.e., sediment being conveyed out of the well mixed layer to depths sufficient that it is essentially inaccessible
(7) flow of water from surface soil to groundwater (recharge)

It also transpires that there are several advective processes which can apply to chemical movement *within* the evaluative environment. Consumption of food containing chemical is another important advective process in bioconcentration. We delay their treatment until later.

In situations 1 to 4, there is no difficulty in deducing the rate as GC or Df, where G is the flowrate of the phase in question, C is the concentration of chemical in that phase, and the Z-value applies to the chemical in the phase in which it is dissolved or sorbed.

For example, aerosol may be transported to an evaluative world in association with inflow of 10^{12} m^3/h of air, the aerosol concentration being 10^{-11} volume fraction. The flowrate of aerosol G_X is then 10 m^3/h. The concentration of chemical is the concentration in, or on, the aerosol, not in the air, and is normally quite high; for example, 100 mol/m^3. Thus, the rate is 1000 mols/h. This can be calculated using the D and f route as follows, giving the same result.

If $Z_X = 10^8$, then $f = C_X/Z_X = 100/10^8 = 10^{-6}$ Pa

$D_{AX} = G_X Z_X = 10 \times 10^8 = 10^9$

Therefore $N = Df = 10^9 \times 10^{-6} = 1000$ mol/h

Treatment of transport to the stratosphere is somewhat more difficult. We can conceive of parcels of air that migrate from the troposphere to the stratosphere at an average, continuous rate, G m^3/h. These are replaced by clean stratospheric air which migrates downward at the same rate. We can thus calculate the D-value. As discussed by Neely and Mackay (1982), this rate should correspond to a residence time of air in the troposphere of about 60 years, i.e., G is V/t. Thus, if V is 6×10^9 and t is 5.25×10^5 h, G is 11400 m^3/h. This rate is very slow and is usually insignificant, but there are two situations in which it is worth including.

We may be interested in calculating the amount of chemical which actually reaches the stratosphere; for example, freons which catalyze the decomposition of ozone. This slow rate is thus important from the viewpoint of the receiving stratospheric phase, but is not particularly important from the delivery, or tropospheric phase. Second, if a chemical is very stable and shows no tendency to be removed from the atmosphere by reaction or deposition processes, then transfer to the stratosphere may be the only mechanism of removal. It then becomes very important as the "only way out." Volatile halogenated hydrocarbons tend to be in this class. If we emit a chemical into the evaluative world at a steady rate by emission and allow for no removal mechanisms whatever, we then encounter considerable difficulty because the concentration will continue to build up indefinitely. Such situations are likely to arise if we view the evaluative world as merely a scaled down version of the entire global environment. There is certainly advective flow of chemical from, for example, the United States to Canada, but there is no advective flow of chemical out of the entire global atmospheric environment, except for the small amounts which transfer to the stratosphere. Whether or not advection is included thus depends upon the system being simulated.

Sediment burial is the process by which chemical is conveyed from the active mixed layer of accessible sediment into inaccessible buried layers. As was discussed earlier, this is a rather naive picture of a complex process, but at least it is a starting point for calculations. The reality is that the sediment is not falling, it is the mixed layer that is rising, eventually filling the lake. Typical burial rates are 1 mm/year, the material being buried being typically 25% solids, 75% water. But as it "moves" to greater depths, water becomes squeezed out. Mathematically, the D-value consists of two terms, the burial rate of solids, and that of water.

For example, if a lake has an area of 10^7 m^2 and has a burial rate of 1 mm/year, the total rate of burial is 10,000 m^3/year or 1.14 m^3/h, consisting of 0.29 m^3/h of solids (G_S) and 0.85 m^3/h of water (G_W). The rate of loss of chemical is then

$$G_S C_S + G_W C_W = G_S Z_S f + G_W Z_W f = f(D_{AS} + D_{AW})$$

Usually, there is a large solid to pore water partition coefficient, thus C_S greatly exceeds C_W, or alternatively Z_S greatly exceeds Z_W, thus the term D_{AS} dominates. A residence time of solids in the mixed layer can be calculated as the volume of solids in the mixed layer divided by G_S. For example, if the depth of the mixed layer is 3 cm, and the solids concentration is 25%, then the volume of solids is 75,000 m^3 and the residence time is 259000 hours, or 29.5 years. The residence time of water is not necessarily the same because the water content is probably higher in the active sediment than in the buried sediment. In reality, the water would exchange diffusively with the overlying water in that time period.

These last three advection processes are quite different in character from the common wind and water flow advective processes; thus, a case can be made for treating them as "pseudo-reactions." We examine this alternative approach later.

Loss of chemical to groundwater by leaching from surface soils is not usually important from the viewpoint of the surface soil since biodegradation, evaporation, and uptake by plants can be much faster. It can be very important as a source of groundwater contamination, especially in agricultural regions where pesticides are applied. The D value can be calculated from the water flowrate G and the water Z value. If necessary, transport of chemical in organic colloids can also be included.

6.3   DEGRADING REACTIONS

Reactions and Nonreactions

The word "reaction" merits definition. We regard reactions as processes which alter the chemical nature of the solute, i.e., change its Chemical Abstract System (CAS) number. For example, hydrolysis of ethyl acetate to ethanol and acetic acid is obviously a reaction, as is conversion of 1,2-dichlorobenzene to 1,3-dichlorobenzene, or even conversion of cis butene-2 to trans butene-2. In contrast, processes which merely convey the chemical from one phase to another, or store it in inaccessible form are not reactions.

Uptake by biota, sorption to suspended material, or even uptake by enzymes are not reactions. A reaction may subsequently occur in these locations, but it is not until the chemical structure is actually changed that we consider "reaction" to have occurred. In the literature, the word reaction is occasionally, and wrongly, applied to these processes, especially to sorption.

We have two tasks. First is the assembly of the necessary mathematical framework for treating reaction rates using rate constants, and second is obtaining information on values of these rate constants.

Reaction Kinetic Algebra

We prefer, whenever possible, to use a simple first order kinetic expression for all reactions. The basic rate equation is

$$\text{rate N} = \text{VCk mol/h}$$

where V is the volume of the phase (m^3), C is the concentration of the chemical (mol/m^3), and k is the first order rate constant with units of reciprocal time. The group VCk thus has units of mol/h.

The classical application of this equation is to radioactive decay, which is usually expressed in the form

$$dC/dt = -Ck$$

integrating from an initial condition of C_O at zero time gives the following equations

$$\ln(C/C_O) = -kt \quad \text{or} \quad C = C_O \exp(-kt)$$

Rate constants have units of frequency or reciprocal time and are thus not easily grasped or remembered. A favorite trick question of examiners is to ask a student to convert a rate constant of 24 h^{-1} into reciprocal days. The correct answer is 576 days^{-1}, so beware of this conversion! It is more convenient to store and remember half-times, i.e., the time, t, which is necessary for C to fall to half of C_O. This can be related to the rate constant as follows

$$\text{when C} = 0.5\ C_O \text{ then t} = t_{1/2}$$
$$\ln (0.5) = -kt_{1/2} \text{ therefore } t_{1/2} = 0.693/k$$

For example, a radio-isotope with a half life of 10 hours, has a rate constant, k, of 0.0693 h^{-1}.

Non-First Order Situations

Unfortunately, there are situations in which the real reaction is not first order, but we can often circumvent this difficulty by expressing it in terms of a pseudo first order rate constant. For example, a microbial decay rate may be proportional to the concentration of microorganisms (e.g., P cells/m^3), and a rate constant applicable to each cell, k_C,

$$\text{thus } N = VPk_CC$$

We can lump Pk_C as a single term, k, recognizing that it depends on P, and thus restore the original form of the equation.

Similarly, a photolytic reaction may depend on radiation intensity, I, photons/s, and on terms quantifying the extent to which these photons are absorbed in the reacting medium. These can also be lumped in a pseudo first order rate constant.

Finally, if a chemical 1 reacts with another chemical 2 which is present at a higher or constant concentration, then the rate expression may be given by the following:

$$N = VC_1C_2k_2 = VC_1k$$
$$\text{where } k = k_2C_2$$

Then we can lump the second order rate constant, k_2, and the concentration of chemical 2 together as a single first order rate constant.

Second order rate expressions usually give rise to messy quadratic equations. Fortunately, most pollutants are present at low concentrations and tend not to react with themselves, so second order expressions are uncommon.

Zero order expressions occasionally occur, in which the rate is proportional to concentration to the power zero and is thus independent of C. It is dangerous to include such expressions in computer programs because the equations can now predict a positive rate of reaction even when there is no chemical present. It is embarrassing when computer models calculate negative concentrations of chemicals.

Our strategy is to use every reasonable excuse to force first order kinetics on systems by lumping parameters in k. The dividends which arise are well worth the effort, because subsequent calculations are much easier.

Perhaps most worrisome are situations in which we treat the kinetics of microbial decay of chemicals. It is possible that, at very low concentrations, there is slower or even no reaction because the required enzyme systems are not "turned on." At very high concentrations, the enzyme may be saturated; thus, the rate of degradation ceases to be controlled by the availability of the chemical, and becomes controlled by the availability of enzyme. In other cases, the rate of conversion may be influenced by the toxicity of the chemical to the organism or by the presence of co-metabolites, chemicals which the enzymes recognize as similar to that of the chemical of interest. Microbiologists have no difficulty conceiving of a multitude of situations in which chemical kinetics become very complicated, and very difficult to predict and express. They seem to obtain a certain perverse delight in finding these situations.

Saturation kinetics is usually treated by the Michaelis-Menten equation which can be derived from first principles, or more simply, by writing down the basic first order equation and multiplying the rate expression by the group shown below.

$$\text{Basic expression } N = VCk$$
$$\text{Group } C_M/(C + C_M)$$
$$\text{Combined expression } N = VCC_Mk/(C + C_M)$$

When C is small compared to C_M, the rate reduces to VCk. When C is large compared to C_M, the rate reduces to VC_Mk which is independent of C and corresponds to the maximum, or zero order rate. The concentration, C_M, thus corresponds to the chemical concentration which gives the maximum rate using the basic expression. When C equals C_M, the rate is half the maximum value. This can be (and usually is) expressed in terms of other rate constants for describing the kinetics of association of the chemical with the enzyme.

The rate expression is then usually written in biochemistry texts in the form

$$N/V = C.v_M/(C + k_M)$$

where v_M is a maximum rate or "velocity" equivalent to kC_M, k_M is equivalent to our C_M, and is viewed as a ratio of rate constants. A somewhat similar expression, the Monod equation, is used to describe cell growth.

If kinetics are not first order, it may be necessary to write the appropriate equations and accept the increased difficulty of solution. A somewhat unethical and cunning alternative is to guess the concentration, calculate the rate N using the non-first order expression, then calculate the pseudo first order rate constant in the expression. For example, if a reaction is second order and C is expected to be about 2 mols/m³, V is 100 m³, and the second order rate constant, k_2, is 0.01 m³/mol.h. Then N equals 4 mols/h. We can set this equal to VCk, then k is 0.02 h^{-1}. Essentially, we have lumped Ck_2 as a first order rate constant. This approach must be used, of course, with extreme caution.

Additivity of Rate Constants

A major advantage of forcing first order kinetic on all reactions is that if a chemical is susceptible to several reactions in the same phase, with rate constants k_A, k_B, k_C, etc., then the total rate constant for reaction is ($k_A + k_B + k_C$), i.e., the rate constants simply add. Another favorite trick of perverse examiners is to inform a student that a chemical reacts by one mechanism with a half-life of 10 hours, and by another mechanism with a half-life of 20 hours, and ask for the total half-life. The correct answer is 6.7 hours, not 30 hours. Half-lives add as reciprocals, not directly.

Level II Algebra, Partition Coefficients

We can now perform certain calculations describing the behavior of chemicals in evaluative environments. The simplest is a Level II equilibrium, steady state reaction situation in which there is no advection and there is a constant inflow of chemical in the form of an emission, as depicted in the second diagram in Figure 6.1. When steady state is reached, there must be an equivalent loss in the form of reaction. Starting from a clean environment, the concentrations would build up until they reach a level at which the rates of degradation or loss equal the total rate of input. We further assume that the phases are in equilibrium, i.e., transfer between them is very rapid; thus, the concentrations are related through partition coefficients, or a common fugacity applies. The equations are as follows

$$E = V_1C_1k_1 + V_2C_2k_2 \text{ etc.} = \Sigma V_iC_ik_i$$

Using partition coefficients

$$E = \Sigma V_iC_WK_{iW}k_i = C_W\Sigma V_iK_{iW}k_i$$

from which C_W can be deduced, followed by other concentrations, amounts, rates of reaction, and the persistence. In these equations a K_{WW} or water-water partition coefficient of unity may be defined.

Worked Example 6.2

The evaluative environment in Example 6.1 is subject to emission of 10 mol/h of chemical, but no advection. The reaction half-lives are air 69.3 hours; water 6.93 hours and sediment 693 hours. Calculate the concentrations. Recall that $K_{AW} = 0.004$ and $K_{SW} = 10$.

The rate constants are 0.693/half-lives or: air 0.01; water 0.1; sediment 0.001; h^{-1}.

$$
\begin{aligned}
E &= V_AC_Ak_A + V_WC_Wk_W + V_SC_Sk_S \\
&= C_W(V_AK_{AW}k_A + V_Wk_W + V_SK_{SW}k_S) \\
&= C_W(0.4 + 10 + 0.01) = C_W(10.41) = 10 \\
\therefore C_W &= 0.9606 \text{ mol/m}^3, \; C_A = 0.0038, \; C_S = 9.606
\end{aligned}
$$

The rates of reaction are then:

air   = 0.384 mol/h
water   = 9.606 mol/h
sediment = 0.010 mol/h

which add to give the emission rate of 10 mol/h.

It is noteworthy that the reaction rate is controlled by the product V, C, and k. A large value of any one of these may convey the wrong impression that the reaction is most important.

Level II Using Fugacity and D Values for Reaction

We can now follow the same process as used when treating advection and define D-values for reactions. If the rate is VCk or VZfk, it is also $D_R f$, where D_R is VZk. Note that D_R has units of mol/m^3.Pa identical to those of D_A or GZ, discussed earlier. If there are several reactions occurring to the same chemical in the same phase, then each reaction can be assigned a D-value, and these D-values can be added to give a total D-value. This is equivalent to adding the rate constants. The Level II mass balance becomes

$$E = \Sigma V_i C_i k_i = \Sigma V_i Z_i fk_i = f\Sigma V_i Z_i k = f\Sigma D_R$$

thus f can be deduced, followed by C, m, rates, and persistence.

We can thus calculate concentrations, amounts, the total amount M, and the rates of individual reactions as VCk or Df. We can repeat Example 6.2 in fugacity format.

air	$V_A = 10^4$	$Z_A = 4 \times 10^{-4}$	$k_A = 0.01$	$D_{RA} = 0.04$
water	$V_W = 100$	$Z_W = 0.1$	$k_W = 0.1$	$D_{RW} = 1.0$
sediment	$V_S = 1$	$Z_S = 1.0$	$k_S = 0.001$	$D_{RS} = 0.001$
				Total = 1.041

$\therefore$ f $= E/\Sigma D = 10/1.041 = 9.606$
$\therefore$ C_A $= 0.0038$ rate $= Df = 0.384$
     C_W $= 0.9606$ Df $= 9.606$
     C_S $= 9.606$ Df $= 0.010$

which is the same result as before.

Another example may be useful.

Worked Example 6.3

An evaluative environment consists of 10000 m^3 air, 100 m^3 water, and 10 m^3 sediment. There is input of 25 mol/h of chemical which reacts with half-lives of 4.17 days in air, 3.125 days in water, and 2.08 days in soil. Calculate the concentrations and amounts given the Z values below:

Phase	volume m^3	Z	k	VZk or D	C mol/m^3	m mol	Rate
Air	10000	4×10^{-4}	0.00693	0.0277	.0386	386	2.68
Water	100	0.1	0.00924	0.0924	9.66	966	8.93
Sediment	10	1.0	0.0139	0.1386	96.6	966	13.39
				0.2587			

The rate constants in each case are 0.693/half-life in hours. The sum of the VZk terms or D values is 0.2587 thus

$$f = E/\Sigma D = 96.6 \text{ Pa.}$$

Thus each C is Zf, the amount m is VC, totaling 2318 mol. Each rate is VCk or Df, totaling 25 mol/h.

It is clear that the D value VZk controls the overall importance of the process. Despite its low volume and relatively slow reaction, the sediment provides a fairly fast reacting medium because of its large Z value. It is not usually immediately obvious where most reaction occurs. The overall residence time is 2318/25 or 93 hours.

Note that the persistence or M/E is a weighted mean of the persistences or of the reciprocal rate constants in each phase. It is also $\Sigma VZ/\Sigma D$.

6.4 COMBINED ADVECTION AND REACTION

Advection and reaction processes can be included in the same calculation as shown in the example below, which is similar to those presented earlier for reaction. We now have inflow and outflow of air and water at rates given below, and with background concentrations. The mass balance equation becomes

$$I = E + G_A C_{BA} + G_W C_{BW} = G_A C_A + G_W C_W + \Sigma V_i C_i k_i$$

This can be solved either by substituting $K_{iW} C_W$ for all concentrations and solving for C_W, or calculating the advective D values as GZ and adding them to the reaction D values. We could illustrate the equivalence of these routes by performing both calculations. The input-output balance is as shown in the third diagram in Figure 6.1.

Worked Example 6.4

The advective flows in Example 6.1 are combined with the environment and reactions of Example 6.3 and with a total emission of 29 mol/h. Calculate the fugacity concentrations, amounts, and chemical residence time.

	Volume	Z	D (advection)	D (reaction)	C	m	rate Df
Air	10000	4×10^{-4}	0.4	0.0277	0.021	210	22.55
Water	100	0.1	0.1	0.0924	5.27	527	10.14
Sediment	10	1.0	0	0.1386	52.7	527	7.31
			0.5	0.2587			40.00

The total of all D values is 0.7587

$$I = E + G_A C_{BA} + G_W C_{BW} = 29 + 10 + 1 = 40$$
therefore $\quad f = 40/\Sigma D = 52.7$ Pa

The total amount is 1264 mols, giving a mean residence time of 31.6 hours. The most important loss process is advection in air, which accounts for 21.08 mol/h; next is sediment reaction at 7.31 mol/h; the water advection at 5.27 mol/h etc.; each rate being Df. The relative importance of each process is clearly indicated by its D value, regardless of whether it is advection or reaction.

Advection as a Pseudo-Reaction

Examination of these equations shows that the advection group G/V plays the same role as a rate constant having identical units of h^{-1}. It may, indeed, be more convenient to regard advective loss as a pseudo-reaction with this rate constant and applicable to the phase volume of V. Note that its reciprocal, the group V/G is the residence time of the phase in the system. Frequently, this is the most accessible and easily remembered quantity. For example, it may be known that the retention time of water in a lake is 10 days, or 240 hours. The advective rate constant, k, is thus $1/240\ h^{-1}$, and the D-value is VZk, which is of course, also GZ.

It is noteworthy that this residence time is not equivalent to a reaction half-time, which is related to the rate constant through the constant 0.693 or ln 2. Residence time is equivalent to $1/k$.

Residence Times

Confusion may arise when calculating the residence time or persistence of a chemical in a system in which advection and reaction occur simultaneously. The overall residence time in Example 6.4 is 31 hours, and is a combination of the advective residence time and the reaction time. The presence of advection does not influence the rate constant of the reaction; thus it cannot affect the persistence of the chemical. But by removing the chemical, it does affect the amount of chemical which is available for reaction, and thus it affects the rate of reaction. It would be useful if we could establish a method of breaking down the overall persistence or residence time into the time attributable to reaction, and the time attributable to advection. This is best done by modifying the fugacity equations as shown below:

$$I = \Sigma D_{Ai}f + \Sigma D_{Ri}f$$

where I is the total input

$$\text{therefore } 1/f = \Sigma D_{Ai}/I + \Sigma D_{Ri}/I$$

but $I = M/t_O$ where M is the amount of chemical and t_O is the overall residence time. Further, $M = \Sigma VZf$ or $f\Sigma VZ$, thus

$$1/t_O = I/M = I/f\Sigma VZ = \Sigma D_{Ai}/\Sigma VZ + \Sigma D_{Ri}/\Sigma VZ$$
$$= 1/t_A + 1/t_R$$

The key point is that the advective and reactive residence times t_A and t_R add as reciprocals to give the reciprocal overall time. Clearly the shorter residence time dominates, corresponding, of course, to the faster rate constant. It can be shown that the ratio of the amounts removed by reaction and by advection are in the ratio of the overall rate constants or the reciprocal times.

Worked Example 6.5

Calculate the individual and overall residence times in Example 6.4. Each residence time is $\Sigma VZ/D$.

	VZ	$\Sigma VZ/D$ (advection)	$\Sigma VZ/D$ (reaction)
Air	4	60	866
Water	10	240	260
Sediment	10	∞	173
	24		

Adding the reciprocals gives

$$1/60 + 1/240 + 1/\infty + 1/866 + 1/260 + 1/173$$
$$= .0167 + 0.0042 + 0 + 0.0012 + 1/0.0038 + .0058$$
$$= 0.0316 = 1/31.6$$

Each residence time (e.g., 60, 866, etc.) contributes to give the overall residence time of 31.6 hours, *reciprocally*. The overall advection residence time is 48 hours and for reaction it is 93 hours.

These concepts are useful because they convey an impression of the relative importance of advective flow (which merely moves the problem from one region to another) versus reaction, which usually solves the problem. These are of particular interest to those who live downwind or downstream of a polluted area.

6.5   UNSTEADY STATE CALCULATIONS

A second calculation can be done in unsteady state mode in which we introduce an amount of chemical, M, into the evaluative environment at zero time, then allow it to decay in concentration with time, but maintain equilibrium between all phases at all times. This is analogous to a batch chemical reaction system. Although it is possible to include emissions or advective inflow, we prefer to treat first the case in which only reaction occurs. The differential equation is:

$$dM/dt = -\Sigma V_i C_i k_i = -f\Sigma V_i Z_i k_i$$
$$\text{But } M = \Sigma V_i Z_i f = f\Sigma V_i Z_i$$
$$df/dt = -f\Sigma V_i Z_i k_i / \Sigma V_i Z_i$$
$$f = f_0 \exp(-k_0 t)$$

where $k_0 = \Sigma V_i Z_i k_i / \Sigma V_i Z_i = \Sigma D_{Ri} / \Sigma V_i Z_i$.

Worked Example 6.6

Calculate the time necessary for the environment in Example 6.3 to recover to 50%, 36.7%, 10%, and 1% of the steady state level of contamination if emissions cease.

Here ΣVZ is 24 and ΣD is 0.2587 thus

$$f = f_0 \exp(-0.2587t/24) = f_0 \exp(-0.01078t)$$

Since M is proportional to f, and f_0 is 96.6 Pa, our aim is to calculate t at which f is 48.3, 35.4, 9.66, and 0.966 Pa. Substituting and rearranging gives $t = -1/0.01078 \ln(48.3/96.6)$ etc., or t is respectively 64 hours, 93 hours, 214 hours, and 427 hours. The 93 hour time is significant as both the steady state residence time and the time of decay to 36.7% or $\exp(-1)$ of the initial concentration.

It is possible to include advection and emissions with only slight complications to the integration, because the input terms are no longer zero.

This example raises an important point which we will address later in more detail. The steady state situations in the Level II calculations are somewhat artificial and contrived. Rarely is the environment at steady state; things are usually getting better or worse. The valid criticism can be leveled at Level II calculations that steady state analysis does not convey information about the rate at which systems will respond to changes. For example, a steady state analysis of salt emission into Lake Superior may demonstrate what the ultimate concentration of salt will be, but it may take 1000 years for this steady state to be achieved. In a much smaller lake, this steady state may be achieved in 10 days. Detractors of steady state models thus point with glee to situations in which the modeler will be dead long before steady state is achieved.

Proponents of steady state models respond that, although they have not specifically treated the unsteady state situation, the equations do contain much of the key "response time" information, which can be extracted with the use of some intelligence. The response time in the unsteady state Example 6.6 was 93 hours, which was $\Sigma VZ/\Sigma D$. This is identical to the overall residence time, t, in Example 6.3. It appears that the response time of an unsteady state system is essentially equivalent to the residence time in a steady state system. By inspection of the magnitude of groups such as VZ/D or the reciprocal rate constants which occur in steady state analysis, it is possible to determine the likely unsteady state behavior. This is bad news to those who enjoy setting up and solving differential equations, because back-of-the-envelope calculations often show that it is not necessary to undertake a complicated unsteady state analysis.

6.6 THE NATURE OF ENVIRONMENTAL REACTIONS

The most important environmental reaction processes are biodegradation, hydrolysis, oxidation, and photolysis. We treat each process briefly below with a view to providing an entry to the literature and to devising experimental and calculation methods by which the rate of the reaction can be characterized. A useful set of experimental techniques has been compiled by the U.S. EPA in its "Chemical Fate Test Guidelines" Report (EPA, 1982). This includes recipes for determining rates of biodegradation, complex formation, hydrolysis, and photolysis.

Biodegradation

Microbiologists are usually quick to point out that the process of microbial conversion of chemicals in the environment is exceedingly complex. The rate of conversion depends on the nature of the chemical compound, on the amount and condition of enzymes which may be present in various organisms in various states of activation and available to perform the chemical conversion, on the availability of nutrients such as nitrogen, phosphorus, and oxygen, as well as temperature and the presence of other substances which may help or hinder the conversion process. Most naturally occurring organic chemicals are susceptible to microbial conversion or biodegradation. The apparent exceptions are high molecular weight compounds such as the humic acids, certain terpenes which appear to have structures which are too difficult for enzymes to attack, and many organo-chlorine chemicals. Generally, water soluble organic chemicals are fairly readily biodegraded. Over evolutionary time, enzymes have apparently adapted and evolved the capability of handling most naturally occurring organic compounds. When presented with certain synthetic organic compounds which do not occur in nature (notably the chlorinated hydrocarbons), they experience considerable difficulty and they may, or may not, be able to effect useful chemical conversions. In such cases, if environmental degradation does take place, it is often the result of abiotic processes such as photolysis or reaction with free radicals.

Our aim is to be able to define a half-life or rate constant for microbial conversion of the chemical, usually in water, but also in soil and in sediments. These rate constants may be measured by introducing the chemical into the medium of interest and following its decay in concentration. If first order behavior is observed, a rate constant and half-life may be established. Care must be taken to ensure that decay is truly attributable to biodegradation and not to other processes such as volatilization.

In many cases non-first order behavior occurs. For example, it is suspected that in some situations the concentration of chemical is so low that the enzymes necessary for conversion do not become adequately activated and the chemical is essentially ignored. At high concentrations, the presence of the chemical may result in toxicity to the microorganisms and thus the conversion process is terminated. The number of active enzymatic sites may also be limited; thus, the rate of conversion of the chemical species becomes controlled, not by the concentration of the chemical but by the number of active sites and the rate at which chemical can be transferred into and out of these sites. Under these conditions of saturation a Michaelis-Menten type equation can be applied as described earlier.

Much to the chagrin of microbiologists we will adopt a simple expedient assuming that a first order rate constant or half-life applies and that the rate constant can be estimated by experiment, or from experience. This is necessarily an inaccurate process and often involves merely a judgment that in this type of water or soil, this compound is subject to biodegradation with a half-life of approximately x hours. The rate constant is therefore $0.693/x$ hours^{-1}. Valiant efforts have been made to devise experimental protocols in which chemicals are subjected to microbial degradation conditions using, for example, inoculated sewage sludge. Such estimates are of particular importance in the prediction of chemical fate in sewage treatment plants. Even more valiant attempts are being made to predict the rate of biodegradation of chemicals purely from a knowledge of their molecular structure. Others have been content to categorize organic chemicals into various groups which have similar biodegradation rates or characteristics.

Several standard and near-standard tests exist for determining biodegradation rates under aerobic and anaerobic conditions in water and in soils. Simplest is the BOD test as described in various standard methods compilations by agencies such as the American Society for Testing and Materials and the American Public Health Association. More complex systems involve the use of chemostats and continuous flow systems which are analogous to bench-top sewage treatment plants.

Some useful references to environmental biodegradation include general accounts of the process by Alexander (1981, 1985), Neilson et al. (1985), Painter and King (1985), and Paris et al. (1982). The degradation of halogenated chemicals is of particular interest as discussed by Brown et al. (1987), Ghosal et al. (1985), and Wood (1987). Methods of estimating or extrapolating rates are reviewed by Johnson (1980), Howard and Banerjee (1984), and Liu et al. (1981), and a kinetic model has been devised by Banerjee et al. (1984). Attempts to correlate degradability with chemical structure have been treated by Wolfe et al. (1980). Databases are described by Howard et al. (1986, 1987). The "low concentration" and related issues are reviewed by Zaidi et al. (1988, a, b; 1989). There is a formidable literature on petroleum degradation, useful entries to the literature being two reviews of marine oil spill fate by the National Academy of Sciences (1975, 1985).

Hydrolysis

In this process the chemical species is subject to addition of water as a result of reaction with water, hydrogen ion, or hydroxyl ion. All three mechanisms may occur simultaneously at different rates; thus, the overall rate can be very sensitive to pH. Rates of environmental hydrolysis have been thoroughly reviewed by Mabey and Mill (1978), who have shown, for example, that one would expect most esters to be converted with the half-life of approximately two days in river water of pH 6. For many organic compounds, hydrolysis is not feasible.

A systematic method of testing for susceptibility to hydrolysis is to subject the chemical to pHs of 3, 7, and 11; observe the decay; and deduce rate constants for acid, neutral, and base hydrolysis. These rate constants can be combined to give an expression for the rate at any desired pH.

Useful references on hydrolysis include the review by Mabey and Mill (1978) and papers by Wolfe (1980), Pankow and Morgan (1981), Zepp et al. (1975), Wolfe et al. (1977) and Jeffers et al. (1989).

Photolysis

The energy present in sunlight photons is often sufficient to cause chemical reactions or rupture of chemical bonds in molecules which are able to absorb this light. Sunburn and photosynthesis are practical examples. This process is primarily of interest when considering the fate of chemicals in solution in water. The radiation which is most likely to effect chemical change is from high energy short wavelength photons at the blue and near UV end of the spectrum, i.e., about 390 nm. If the absorption characteristics of the molecules are such that the photon is absorbed, then the molecule may become activated and dissociate or react, forming another chemical species. In principle, the calculation involves an estimation of the number of photons absorbed by the solution, bearing in mind that other chemical species including, for example, fulvic and humic acids, will be absorbing light in competition, and the calculation of the efficiency with which the activated species reacts, i.e., the quantum yield. In practice, this calculation is quite difficult because as light penetrates into a water body it is attenuated by absorption by water, and by material dissolved and suspended in the water. Thus the photolytic activity decreases with depth in the water column and a calculation must be made to correct for this absorption. The rate is, of course, also influenced by the amount of radiation reaching the water surface, as influenced by time of day, time of year, and latitude and cloud cover.

Relatively simple experiments can be conducted in which the chemical is dissolved in distilled or natural water in a suitable container and exposed to natural sunlight or to an artificial lamp for a period of time, and the concentration decay monitored. Procedures have been developed by which photolysis rates can be calculated for species in ponds and lakes. These calculations have been incorporated into environmental fate models such as the U.S. EPA EXAMS model.

The issue is complicated by the presence of photosensitizing molecules or substances. These substances absorb light, then pass on the energy to the chemical of interest, resulting in subsequent chemical reaction. It is therefore not necessary for the chemical to absorb the photon directly. It can receive it "second hand" from a photosensitizer. This is a troublesome complication because it raises the possibility that chemicals may be subject to photolysis due to the unexpected presence of a photosensitizer. Of particular interest are the naturally occurring organic matter photosensitizers which are present in water and give it its characteristic brown color, especially in areas in which there is peat and decaying vegetation.

A pioneering study of photolysis is the work of Zepp and Cline (1977). Related papers are by Zepp and coworkers (1975, 1977, 1978, 1980, 1982, 1985). Wong and Crosby (1981), Miille and Crosby (1983), Roof (1982), Zafiriou et al. (1984), and Payne and Phillips (1985) have described and discussed marine and freshwater photolysis. Test methods have been described by Svenson and Bjorndal (1988), Lemaire et al. (1982), and Dulin and Mill (1982). Photosensitization has been discussed by Miller and Zepp (1979, 1983), Mudambi and Hassett (1988), van Noort et al. (1988) and Faust and Holgne (1987). Several texts on air pollution chemistry provide details of the complex sunlight-initiated processes which occur to atmospheric pollutants, e.g. Seinfeld (1975).

Oxidation Reactions

The chemical may react with oxygen, an activated form of oxygen such as singlet oxygen, with ozone, with hydrogen peroxide, or with various free radicals; notably OH radicals. Fortunately we live in a world with an abundance of oxygen, and it is not surprising that a suite of oxygen compounds exists eager to oxidize organic chemicals. The rates of these reactions can be estimated by conducting conventional chemical kinetic experiments in which the substance is contacted with known concentrations of the oxidant, the decay of chemical is followed, and a kinetic law and rate constant established.

One of the most important oxidative processes is the reaction of hydroxyl radicals with chemical species in the atmosphere. The concentration of these hydroxyl radicals is exceedingly small, probably only about a million molecules per cubic centimeter, but they are extremely reactive and are responsible for the reaction of many organic chemicals in the environment which would otherwise be very persistent. Readers should consult books on atmospheric chemistry or air pollution for more authoritative accounts of atmospheric reaction processes. It is important to appreciate that the atmosphere is a very reactive medium in which large quantities of chemical species are converted into oxidized products. This is fortunate, because otherwise there would be severe air pollution and problems associated with the transport of these chemicals to remote regions.

Other Reactions

Chemicals may be susceptible to reaction with other species in the environment; for example, chlorine or ozone introduced for disinfectant purposes, and naturally occurring acids and alkalis.

Summary

It has only been possible to provide a brief account of the vast literature relating to chemical reactivity in the environment. The air pollution literature is particularly large and detailed. References have been provided, not to give a complete coverage, but merely to give the reader an entry to the literature.

The susceptibility of a chemical in a specific medium to degrading reaction depends both on the inherent properties of the molecule and on the nature of the medium, especially temperature and the presence of candidate reacting molecules or enzymes. In this respect environmental chemicals are fundamentally different from radio-isotopes, which are totally unconcerned about their medium. Translation and extrapolation of reaction rates from environment to environment and laboratory to environment is thus a challenging and fascinating task which will undoubtedly keep environmental chemists busy for many more decades.

6.7   LEVEL II CALCULATIONS

As with Level I calculations, it is desirable to systematize calculations using a calculation form, and to reduce the tedium of calculations by using the computer. Figure 6.2 gives an illustrative fugacity form calculation, a blank form being provided in the appendix. Computer programs which conduct Level II calculations are provided on the diskette. The input data include the properties of the environment, the chemical properties, input rates by emission and advection, reaction and advection rates. The fugacity is calculated, followed by a complete mass balance. Since equilibrium is assumed to apply within the environment, it is immaterial into which medium the chemical is introduced. The program and output are self explanatory.

The user is encouraged to test the environmental behavior of some of the chemicals introduced earlier, assuming or obtaining literature data on reaction rates.

FUGACITY FORM 3			LEVEL II

CHEMICAL: HYPOTHENE

Direct emission rate E 100 mol/h

Advective input rates

Compartment	AIR	WATER		
Volume m³ (V)	6×10^9	7×10^6		
Residence time h (t)	600	7000		
Flow rate m³/h = V/t = G	10^7	1000		
Inflow concentration mol/m³ C_B	10^{-6}	10^{-2}		
Chemical inflow rate mol/h = GC_B	10	10		

Total input rate $E + \Sigma GC_B = I = $ $100 + 10 + 10 = 120$

Compartment	AIR	WATER	SOIL	SEDIMENT
Volume V m³	6×10^9	7×10^6	45000	21000
Z	4×10^{-4}	0.1	12.3	24.6
VZ	2.4×10^6	7×10^5	5.5×10^5	5.17×10^5
Reaction half life (h) t	∞	643	69.3	6930
Rate constant k = 0.693/t (h⁻¹)	0	0.001	0.01	0.0001
Advective flow G m³/h	10^7	1000	0	0
D reaction = VZk = D_R	0	700	5540	51.7
D advection = GZ = D_A	4000	100	0	0
$D_R + D_A = D_T$	4000	800	5540	51.7

Total D value = $\Sigma D_T = $ 10392 Fugacity f = $1/\Sigma D$ = $120/10392 = 1.15 \times 10^{-2}$

	AIR	WATER	SOIL	SEDIMENT
C = Zf mol/m³	4.6×10^{-6}	1.15×10^{-3}	0.14	0.28
m = CV mol	2.8×10^4	8050	6400	5900
percent	57.5	16.8	13.3	12.4
C_G g/m³, i.e. CW	9.2×10^{-4}	0.23	28	56
Density ρ kg/m³	1.17	1000	1500	1500
C_U μg/g, i.e. $C_G \times 1000/\rho$	0.78	0.23	19	38
Reaction rate $D_R f$	0	8.1	63.9	0.6
Advection rate $D_A f$	46.2	1.2	0	0
Total rate $D_T f$	46.2	9.3	63.9	0.6

Total amount M = $\Sigma m = $ 48300

Total reaction rate = $\Sigma D_R f = $ 72.6 Reaction residence time (h) = $M/\Sigma D_R f = $ 665

Total advection rate = $\Sigma D_A f = $ 47.4 Advection residence time (h) = $M/\Sigma D_A f = $ 1018

Total output rate (mol/h) = I = 120 Overall residence time (h) = M/I = 402

Figure 6.2.   Level II fugacity form.

Worked Example 6.7

Calculate the partitioning of the hypothetical chemical in Example 5.11, assuming rate constants for reaction of 0.001 h^{-1} in water, 0.01 h^{-1} in soil, and 0.0001 h^{-1} in sediment and no reaction in air. Assume advective inputs in air at 10^{-6} mol/m^3 (flow 10^7 m^3/h) and in water at 0.01 mol/m^3 (flow 1000 m^3/h). The emission rate is 100 mol/h.

The hand calculation is fairly tedious and is reproduced in Figure 6.2. It involves calculation of the total inputs of 100 mol/h (emission), 10 mol/h (advection in air), and 10 mol/h (advection in water) totaling 120 mol/h (I). The reaction and advection D values are then deduced and added to give a total (ΣD) of 10392 mol/Pa.h.

The fugacity is then I/ΣD or 0.0115 Pa.

Concentrations, amounts, and process rates can then be deduced and added to check the mass balance.

The computed output is given in Figure 6.3.

FUGACITY LEVEL II CALCULATION

Properties of HYPOTHENE

Temperature deg C	27.5
Molecular mass g/mol	200
Vapor pressure Pa	1
Solubility g/m^3	20
Solubility mol/m^3	.1
Henry's law constant Pa.m^3/mol	10
Log octanol-water p-coefficient	4
Octanol-water partn-coefficient	10000
Organic C-water ptn-coefficient	4100
Air-water partition coefficient	4.000 E-03
Soil-water partition coefficient	123
Sedt-water partition coefficient	246
Emission rate of chemical mol/h	100
Fugacity Pa	1.155 E-02
Total of VZ products	4170480
Total amount of chemical mol	48179

Phase properties, compositions and rates

Phase	Air	Water	Soil	Sediment
Volume m^3	6E+09	7000000	45000	21000
Density kg/m^3	1.175	1000	1500	1500

(continued on next page)

(continued from previous page)

Frn org carb	0	0	.02	.04
Z mol/m³.Pa	4.000 E-04	.1	12.3	24.6
Adv. flow m³/h	1E+07	1000	0	0
Adv. resid. time h	600	7000	0	0
Conc in mol/m³	.000001	.01	0	0
Inflow mol/h	10	10	0	0
Rct halflife h	1E+11	693	69.3	6930
Rct rate c.h-1	6.930 E-12	.001	.01	.0001
VZ mol/Pa	2400380	700000	553500	516600
Fugacity Pa	1.155 E-02	1.155 E-02	1.155 E-02	1.155 E-02
Conc mol/m³	4.621 E-06	1.155 E-03	.1420	.2841
Conc g/m³	9.243 E-04	.2310	28.41	56.83
Conc μg/g	.786	.2310	18.94	37.89
Amount mol	27730.58	8086.80	6394.35	5968.06
Amount %	57.55	16.78	13.27	12.38
D rct mol/Pa.h	1.663 E-05	700	5535	51.66
D adv mol/Pa.h	4000.634	100	0	0
Rct rate mol/h	1.921 E-07	8.086	63.94	.596
Adv rate mol/h	46.21	1.155	0	0
Reaction %	1.60 E-07	6.73	53.28	.49
Advection %	38.51	.96	0	0

Total advection D value		4100.63
Total reaction D value		6286.66
Total D value		10387.29
Total chemical input	mol/h	120
Total chemical output	mol/h	120
Output by reaction	mol/h	72.62
Output by advection	mol/h	47.37
Overall residence time	h	401.49
Reaction residence time	h	663.38
Advection residence time	h	1017.03

Figure 6.3. Specimen output of a Level II computer program.

6.8 SUMMARY

In this chapter, we have developed the capability of quantifying advection and reaction rates in evaluative Level II calculations. These calculations can be done using concentrations and partition coefficients or using fugacities and D-values. The concept of residence time or persistence has been introduced.

It is clear that these terms are invaluable descriptors of environmental fate. We have briefly reviewed the environmental chemistry of biodegradation, photolysis, hydrolysis, and other reactions, and provided references to workers who have measured or calculated the rate constants.

However, critics will be eager to point out a major weakness in these calculations. Rarely are environmental media in equilibrium; thus, the use of a common fugacity, or of equilibrium partition coefficients, to calculate concentrations may not be valid. Treating nonequilibrium situations is the task of Chapter 7.

7 INTERMEDIA TRANSPORT

7.1 INTRODUCTION

The Level II calculations described in Chapter 6 contain a major weakness in that they assume environmental media to be in equilibrium. This is rarely the case in the real environment; thus, the use of a common fugacity (or concentrations related by equilibrium partition coefficients) is often invalid. Reasons for this are best illustrated by an example.

Suppose we have a two medium system as illustrated in Figure 7.1, consisting of air and water, with emissions of 100 mol/h of benzene into the water. There is, we assume, only slow reaction in the water (say 20 mol/h), but there is rapid reaction (say 80 mol/h) in the air. Benzene must therefore evaporate from water to air at a rate of 80 mol/h. The question arises: is benzene capable of evaporating at this rate, or will there be a resistance to transfer which prevents evaporation at this rate? If only 40 mol/h could evaporate, the evaporated benzene may react in the air phase at 40 mol/h, but it will tend to build up in the water phase to a higher concentration or fugacity until the rate of reaction in the water is 60 mol/h. The benzene fugacity in the air will thus be lower than the fugacity in water, and a nonequilibrium situation has developed. We cannot presently tackle this problem because we do not yet have the ability to calculate how fast chemicals can migrate from one phase to another. This is essentially the task of this chapter. It proves to be a challenging task, and one in which there still remains considerable doubt and scope for scientific investigation and innovation. We approach it by first listing and grouping all the transport processes which are likely to occur.

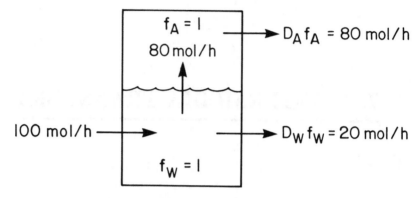

$$\text{EQUILIBRIUM i.e. } f_W = f_A$$
$$D_A = 80, D_W = 20 \text{ mol/Pa.h}$$

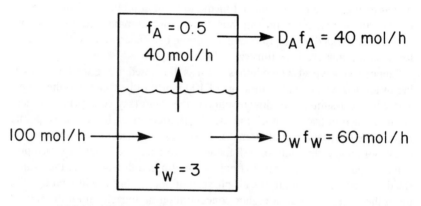

$$\text{NON-EQUILIBRIUM i.e. } f_W > f_A$$

Figure 7.1. Illustration of nonequilibrium behavior in an air-water system.

7.2   DIFFUSIVE AND NONDIFFUSIVE PROCESSES

Nondiffusive Processes

The first group of processes consists of *nondiffusive*, or ''piggyback,'' or advective processes. A chemical may move from one phase to another by "piggybacking" on material which has decided, for reasons unrelated to the presence of the chemical, to make this journey. Examples are advective flow in air, water, or particulate phases, as discussed in Chapter 6, deposition of chemical in rainfall moving from the atmosphere to soil or water, deposition of chemical in aerosol form by wet and dry deposition, transfer of chemical from water to air in spray from breaking waves, sedimentation of chemical in association with particles which fall from the water column to the bottom sediments, runoff of water (and soil associated with that water) from the soil compartment to groundwater and surface water, resuspension of bottom sediment material back to the water column, ingestion of food containing chemical by organisms, subsequent excretion in feces, movement of chemical within organisms by blood flow or, in the case of plants, by flow in the sap.

These are usually one-way processes. The rate of chemical transfer is simply the product of the concentration C mol/m^3 of chemical in the moving medium, and the flowrate of that medium, G, m^3/h. We can thus treat all these processes as advection, and calculate the rate as follows

$$N = GC = GZf = Df \text{ mol/h}$$

The usual problem is to measure or estimate G and the corresponding Z-value or partition coefficient. We examine these rates in more detail later when we focus on individual intermedia transfer processes.

Diffusive Processes

The second group of processes are *diffusive* in nature. Using the transfer of benzene between water and octanol phases as an example, if we have water containing 1 mol/m^3 of benzene and add an octanol phase to it, the benzene will diffuse from the water to the octanol until it reaches a concentration in octanol which is K_{OW}, or 135, times that in the water. We could rephrase this by stating that initially, the fugacity of benzene in the water was, say, 500 Pa, and that in the octanol was zero. The benzene then migrates from water to octanol until the fugacity settles out at a common value of, say, 200 Pa. At this common fugacity, the ratio C_O/C_W is, of course, Z_O/Z_W or K_{OW}. We argue that diffusion will always occur from high fugacity (for example, f_W in water) to low fugacity (f_O in octanol). Thus, it is tempting to write the transfer rate equation from water to octanol as

$$N = D(f_W - f_O)$$

This equation has the correct property that when f_W and f_O are equal, there is no net diffusion. It also correctly predicts the direction of diffusion.

In reality, when the fugacities are equal, there is still active diffusion between octanol and water. Benzene molecules in the water phase do not know what the fugacity is in the octanol phase. At equilibrium, they diffuse at a rate, Df_W, from water to benzene, and this is balanced by an equal rate, Df_O, from octanol to water, i.e., the escaping tendencies have become equal. The term $(f_W - f_O)$ is called the "departure from equilibrium" group.

Other areas of science provide good precedents for using this approach. Ohm's law states that current flows at a rate proportional to voltage difference times electrical conductivity. Electricians prefer to use resistance, which is simply the reciprocal of conductivity. The rate of heat transfer is expressed by Fourier's law as a thermal conductivity times a difference in temperature. Again, it is occasionally convenient to think in terms of a thermal resistance (the reciprocal of thermal conductivity), especially when buying insulation. Another more exotic form of this equation is Newton's law of momentum transfer in which momentum diffuses from regions of high to low velocity. The conductivity term is the viscosity. All these equations have the general form:

$$\text{Rate} = (\text{Conductivity}) \times (\text{Departure from Equilibrium})$$

or $$= (\text{Departure from Equilibrium})/(\text{Resistance})$$

Our task is to devise recipes for calculating D values as an expression of conductivity or reciprocal resistance for a number of processes involving diffusive interphase transfer. These include the following:

1. Evaporation of chemical from water to air and the reverse process of absorption. Note that we consider the chemical to be in solution in water and not present as a film or oil slick, or in sorbed form.
2. Sorption from water to suspended matter in the water column, and the reverse desorption.
3. Sorption from the air to aerosol particles, and the reverse desorption.
4. Sorption of chemical from water to bottom sediment, and the reverse desorption.
5. Diffusion within soils, and from soil to air.
6. Absorption of chemical by fish and other organisms by diffusion through the gills, following the same route traveled by oxygen.
7. Transfer of chemical across other membranes in organisms; for example, from air through lung surfaces to blood, or from gut contents to blood through the walls of the gastrointestinal tract, or from blood to organs in the body.

Armed with these D values we can then set up mass balance equations which are similar to the Level II calculations in Chapter 6, but allow for unequal fugacities between media.

To address these tasks, we return to first principles, quantify diffusion processes in a single phase, then extend this capability to increasingly complex situations involving two or more phases. Chemical engineers have discovered that it is possible to make a great deal of money by inducing chemicals to diffuse from one phase to another. Examples are the separation of alcohol from fermented liquors to make spirits, the separation of gasoline from crude oil, the removal of salt from sea water, and the removal of metals from solutions of dissolved ores. They have thus devoted considerable effort to quantifying diffusion rates, and especially to accomplishing diffusion processes inexpensively in chemical plants. We therefore exploit this body of profit-oriented information for the nobler purpose of environmental betterment.

7.3   MOLECULAR DIFFUSION WITHIN A PHASE

In liquids and gases, molecules are in a continuous state of relative motion. If a group of molecules in a particular location is labeled at a point in time, as shown in Figure 7.2, then at some time later it will be observed that they have distributed themselves randomly throughout the available volume of fluid.

If the number of molecules is large (and it usually is), it is exceedingly unlikely that they will ever return to their initial condition. This process is merely a manifestation of mixing in which one specific distribution of molecules gives way to one of many other statistically more likely mixed distributions. This phenomenon is easily demonstrated by combining particles of two colors, then shaking them to obtain a homogeneous mixture. It is the rate of this mixing process which is at issue.

We approach this issue from two points of view. First is a purely mathematical approach, in which we postulate an equation that describes this mixing, or diffusion, process. Second is a more fundamental approach in which we seek to understand the basic determinants of diffusion in terms of molecular velocities.

Most texts follow the mathematical approach and introduce a quantity, termed diffusivity or diffusion coefficient, which has dimensions of m^2/h, to characterize this process. It appears as the proportionality constant in the equation expressing Fick's First Law of diffusion, namely

$$N = -B.A \ dC/dy$$

Here, N is the flux of material (mol/h), B is the diffusivity (m^2/h), A is area (m^2), C is concentration of the diffusing material (for example, benzene in

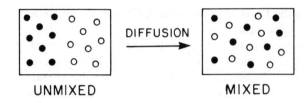

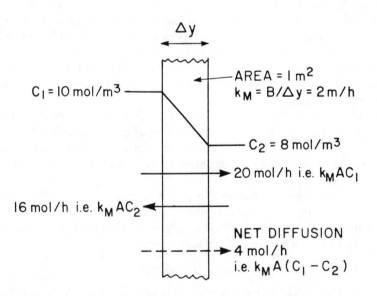

Figure 7.2.   The fundamental nature of molecular diffusion.

water) (mol/m³), and y is distance (m) in the direction of diffusion. The group dC/dy is thus the concentration gradient and is characteristic of the degree to which the solution is unmixed or heterogeneous. The negative sign arises because the direction of diffusion is from high to low concentration, i.e., it is positive when dC/dy is negative. Here, we use the symbol B for diffusivity to avoid confusion with D-values. Most texts sensibly use the symbol D. The equation is really a statement that the rate of diffusion is proportional to the concentration gradient and the proportionality constant is the diffusivity. When the equation is apparently not obeyed, we attribute this misbehavior to deviations or changes in the diffusivity, not to failure of the equation.

As was discussed in Chapter 2.6, there are differences of opinion about the word "flux." We use it here to denote a transfer rate in units such as mol/h. Others insist that it should be area-specific and have units of mol/m.²h.

It is worthwhile digressing to examine the fundamental mixing process which yields this equation. This elucidates the nature of diffusivity and the reason for its rather strange units of m²/h. Much of the pioneering work in this area was done by Einstein in the early part of this century and arose from an interest in Brownian movement; the erratic, slow, but observable motion of microscopic solid particles in liquids which is believed to be due to multiple collisions with liquid molecules.

Fick's Law and Diffusivity

We consider a square tunnel of area A m² containing a nonuniform solution, as shown in Figure 7.2, having volumes V_1, V_2, etc., separated by planes 1-2, 2-3, 3-4, etc., each y m apart.

We assume that the solution consists of identical dissolved particles which move erratically, but on the average travel a horizontal distance y m in t hours. In time t half the particles in volume V_3 will cross the plane 2-3, and half the plane 3-4. They will be replaced by (different) numbers which enter volume V_3 by crossing these planes in the opposite direction from volumes V_2 and V_4. Let the concentration of particles in V_3 and V_4 be C_3 and C_4 mol/m³ so that C_3 exceeds C_4. The net transfer across plane 3-4 will be the sum of the two processes, C_3 yA/2 moles from left to right, and C_4 yA/2 moles from right to left. The net amount transferred in time t is then

$$C_3 yA/2 - C_4 yA/2 = (C_3 - C_4) yA/2 \text{ mol}$$

Note that CyA is the product of concentration and volume, and is thus an amount (moles).

The concentration gradient which is causing this net diffusion from left to right is $(C_3 - C_4)/y$, or in differential form dC/dy. The negative sign below is necessary because C decreases in the direction in which y increases. It follows that

$$(C_3 - C_4) = -ydC/dy$$

The flux or diffusion rate is then N or

$$N = (C_3 - C_4) \ yA/2t = - (y^2A/2t) \ dC/dy \ mol/s = -BAdC/dy \ mol/h$$

which is referred to as Fick's First Law. The diffusivity B is thus $(y^2/2t)$ where y is the molecular displacement which occurs in time t. The diffusion distance y is thus $\sqrt{2Bt}$.

In a typical gas at atmospheric pressure, the molecules are moving at a velocity of some 500 m/s, but they collide after traveling only some 10^{-7} m, i.e., after $10^{-7}/500$ or 2×10^{-10} s. It can be argued that y is 10^{-7} m and t is 2×10^{-10}; thus, we expect a diffusivity of approximately 0.25×10^{-4} m²/s or 0.25 cm²/s, or 0.1 m²/h which is borne out experimentally. The kinetic theory of gases can be used to calculate B theoretically, but more usefully, the theory gives a suggested structure for equations which can be used to correlate diffusivity as a function of molecular properties, temperature, and pressure.

In liquids, the molecular motion is more restricted, collisions occur almost every molecular diameter, and the friction experienced by a molecule as it attempts to "slide" between adjacent molecules becomes important. This frictional resistance is related to the liquid viscosity μ(Pa.s). It can be shown that, for a liquid, the group $(B\mu/T)$ should be relatively constant and (by the Stokes-Einstein equation) approximately equal to $(R/6\pi Nr)$ where N is Avogadro's number, R is the gas constant (8.314 Pa m³/mol K), and r is the molecular radius (typically 10^{-10} m). The viscosity of water μ is typically 10^{-3} Pa.s; thus, since $(R/6\pi Nr)$ is approximately 7×10^{-15} Pa m²/mol K, B is expected to be approximately 2×10^{-9} m²/s or 2×10^{-5} cm²/s or 7×10^{-6} m²/h, which is also borne out experimentally. Again, this equation forms the foundation of correlation equations.

The important conclusion is that, during its diffusion journey, a molecule does not move with a constant velocity related to the molecular velocity. On the average, it spends as much time moving backward as forward, thus its net progress in one direction in a given time interval is not simply velocity/time. In t seconds, the distance traveled (y) will be $\sqrt{2Bt}$ m. Taking typical gas and liquid diffusivities of 0.25×10^{-4} m²/s and 2×10^{-9} m²/s respectively, a molecule will travel in one second distances of 7 mm in a gas and 0.06 mm in a liquid. To double these distances will require four seconds, not two seconds. It thus may take a considerable time for a molecule to diffuse a "long" distance, since the time taken is proportional to the square of the distance. The most significant environmental implication is that for a molecule to diffuse through, for example, a 1 m depth of still water requires (in principle) a time of the order of 3,000 days. A layer of still water 1 m deep can thus effectively act as an impermeable barrier to chemical movement. In practice, of course, it is unlikely that the water could remain still for such a period of time.

The reader who is interested in a fuller account of molecular diffusion is referred to the texts by Reid et al. (1987), Sherwood et al. (1975), Thibodeaux (1979), and Bird et al. (1960).

Mass Transfer Coefficients

Diffusivity is a quantity with some characteristics of a velocity, but dimensionally it is the product of velocity and the distance to which that velocity applies. In many environmental situations, B is not known accurately, nor is y or Δy; thus, the flux equation in finite difference form contains two unknowns, B and Δy.

$$N = -AB\Delta C/\Delta y \text{ mol/h}$$

Combining B and Δy in one term k_M equal to $B/\Delta y$ with dimensions of velocity thus appears to decrease our ignorance, since we now do not know one quantity instead of two. Hence we write

$$N = -Ak_M\Delta C \text{ mol/h}$$

k_M is termed a "mass transfer coefficient," has units of velocity (m/h), and is widely used in environmental transport equations. It can be viewed as the net diffusion velocity. The flux N is then the product of the velocity, area, and concentration.

$$\text{i.e., } N = -k_M AC \text{ mol/h}$$

For example, if, as in Figure 7.2, diffusion is occurring in an area of 1 m^2 from point 1 to 2, C_1 is 10 mol/m^3, C_2 is 8 mol/m^3, and k_M is 2.0 m/h, we may have diffusion from 1 to 2 at a velocity of 2.0 m/h, giving a flux of $k_M AC_1$ of 20 mol/h. There is an opposing flux from 2 to 1 of $k_M AC_2$ or 16 mol/h. The net flux is thus the difference, namely 4 mol/h from 1 to 2, which, of course, equals $k_M A(C_1 - C_2)$.

Fugacity Format, D Values for Diffusion

In summary, we calculate diffusion fluxes N as $-ABdC/dy$ or $AB\Delta C/\Delta y$ or $k_M A\Delta C$. In fugacity format we substitute Zf for C and define D values as $BAZ/\Delta y$ or $k_M AZ$, and the flux is then $D\Delta f$ since ΔC is $Z\Delta f$. Note that the units of D are mol/Pa.h, identical to advection and reaction D values.

$$D = BAZ/\Delta y \text{ or } D = k_MAZ$$
$$N = Df_1 - Df_2 = D(f_1 - f_2)$$

Worked Example 7.1

A chemical is diffusing through a layer of still water 1 mm thick, with an area of 20 m^2 and with concentrations on either side of 10 and 5 mol/m^3. If the diffusivity is 10^{-5} cm^2/s, what is the flux and the mass transfer coefficient?

$$y \text{ is } 10^{-3} \text{ m, B is } 10^{-5} \times 10^{-4} = 10^{-9} \text{ m}^2/\text{s}$$
$$\text{thus, } k_M \text{ is } B/\Delta y = 10^{-6} \text{ m/s}$$

The flux N is thus

$$k_MA(C_1 - C_2) = 10^{-6}.20(10 - 5) = 10^{-4} \text{ mol/s}$$

This flux of 10^{-4} mol/s can be regarded as a net flux consisting of k_MAC_1 or 2×10^{-4} mol/s in one direction and k_MAC_2 or 10^{-4} mol/s in the opposing direction.

Worked Example 7.2

Water is evaporating from a pan of 1 square meter area containing 1 cm depth of water. The rate of evaporation is controlled by diffusion through the thin air film 2 mm thick immediately above the water surface. The concentration of water in the air immediately at the surface is 25 g/m^3 (this having been deduced from the water vapor pressure), and in the room the bulk air contains 10 g/m^3. If the diffusivity is 0.25 cm^2/s, how long will the water take to evaporate completely?

B is 0.25 cm^2/s or 0.09 m^2/h
Δy is 0.002 m
ΔC is 15 g/m^3
$N = AB\Delta C/\Delta y = 675$ g/h
To evaporate 10000 g will take 14.8 hours

Note that the "amount" unit in N and C need not be moles. It can be another quantity such as grams, but it must be consistent in both. In this example the 2 mm is controlled by the air speed over the pan. Increasing the air speed would reduce this to, say, 1 mm, thus doubling the evaporation rate. This Δy is rather suspect, so it is more honest to use a mass transfer coefficient, which in the example above is 0.09/0.002 or 45 m/h. This is the actual net velocity with which water molecules migrate from the water surface into the air phase.

Sources of Molecular Diffusivities

Many handbooks contain compilations of molecular diffusivities. The text *Properties of Gases and Liquids,* by Reid et al. (1987) contains data and correlations as does *Mass Transfer,* by Sherwood et al. (1975). The text by Lyman et al. (1982) gives correlations from an environmental perspective. The correlations for gas diffusivity are based on kinetic theory, while those for liquids are based on the Stokes-Einstein equation. In most cases only approximate values are needed. In some equations the diffusivities are expressed in dimensionless form as Schmidt Numbers Sc where

$$Sc = \mu/\rho B$$

where μ is viscosity and ρ is density.

7.4 TURBULENT OR EDDY DIFFUSION WITHIN A PHASE

So far, we have assumed that diffusion is entirely due to random molecular motion and that the medium in which diffusion occurs is immobile or stagnant with no currents or eddies. In practice, of course, the environment is rarely stagnant, there being currents and eddies induced by the motion of wind and water and, indeed, by biota such as fish or worms. This turbulent motion illustrated in Figure 7.3 also promotes mixing by conveying an element or eddy of fluid from one region to another. The eddies may vary in size from millimeters to kilometers, and a large eddy may contain a fine structure of small eddies. Intuitively, it is unreasonable for an eddy to penetrate an interface; thus, in regions close to interfaces, eddies tend to be damped, and only slippage parallel to the interface is possible. There may, therefore, be a thin layer of relatively quiescent fluid close to the interface which can be referred to as a laminar sublayer. In this layer, movement of solute to and from the interface may be only by molecular diffusion.

Under certain conditions, eddies in fluids may be severely damped or their generation may be prevented. This occurs in a layer of air or water when the fluid density decreases with increasing height. This may be due to the upper layers being warmer or, in the case of sea water, less saline. An eddy which is attempting to move upward immediately finds itself entering a less dense fluid and experiences a hydrostatic "sinking" force. Conversely, a companion eddy moving downward experiences a "floating" force, which also tends to restore it to its original position. This inherent resistance to eddy movement damps out most fluid movement and stable, stagnant conditions prevail. Thermoclines in water, and inversions in the atmosphere are examples of this phenomenon. These stagnant or near-stagnant layers may act as diffusion barriers in which only molecular diffusion or slight eddy diffusion can occur.

Conversely, situations in which density increases with height tend to be unstable, and eddy movement is enhanced and accelerated by the density field.

An attractive approach is to postulate the existence of an "eddy diffusivity," B_E, which is defined identically to the molecular diffusivity, B_M. The flux equation then becomes

$$N = -A(B_M + B_E)dC/dy$$

The task is then to devise methods of estimating B_E for various environmental conditions. We expect that, in many situations, such as in winds or fast rivers, B_E is much greater than B_M, and the molecular processes can be ignored. In stagnant regions, such as thermoclines or in deep sediments, B_E may be small or zero, and B_M dominates. As we move closer to a phase boundary, B_E tends to become smaller, thus it is possible that much of the resistance to diffusion lies in the layer close to the interface. It is likely that the roughness of the interface plays a role in determining the thickness of this layer. For example, grass may damp out wind eddies.

A complicating factor is that we have no guarantee that B_E is "isotropic," i.e., that the same value applies vertically and horizontally. In Figure 7.3, we postulate that some eddies may be constrained to form elongated "roll cells." The horizontal B_E will therefore exceed the vertical value. In practice, this nonisotropic situation is common and even leads to conditions in rivers where three B_E values must be considered: vertical, upstream-downstream, and cross-stream.

Turbulent processes in the environment are thus intensely complex and difficult to describe mathematically. The interested reader can consult Thibodeaux (1979) or Csanady (1973) for a review of the mathematical approaches adopted. We sidestep this issue here, but certain generalizations which emerge from the study of turbulent diffusion are worth noting.

In the bulk of most fluid masses (air and water) which are in motion, turbulent diffusion dominates. We can measure and correlate these diffusivities. Generally, vertical diffusion is slower than horizontal diffusion. Often diffusion is so fast that near-homogeneous conditions exist.

At phase interfaces (e.g., air-water, water-bottom sediment), turbulent diffusion is severely damped or is eliminated; thus, only molecular diffusion remains. One can even postulate the presence of a "stagnant layer" in which only molecular diffusion occurs, and calculate its diffusion resistance. This model is usually inherently wrong in that no such layer exists. It is more honest (and less trouble) to avoid the use of diffusivities and stagnant layer thicknesses close to the phase interfaces and invoke mass transfer coefficients which

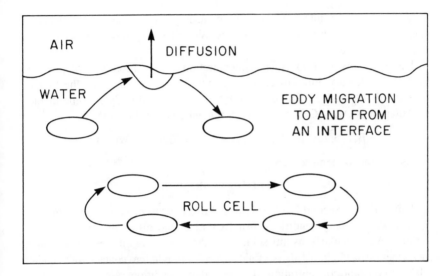

Figure 7.3. The nature of turbulent diffusion.

combine the varying eddy diffusivities, the molecular diffusivity, and some unknown layer thickness, into one parameter, k_M. We then measure and correlate k_M as a function of fluid conditions (e.g., wind speed) but seeking advice from the turbulent transport theorists as to the best form of the correlation equations.

In some diffusion situations, such as bottom sediments, the eddy diffusion may be induced by burrowing worms or creatures which "pump" water. This is termed "bioturbation" and is difficult to quantify. Its high variability and unpredictability is a source of delight to biologists and irritation to physical scientists.

7.5 UNSTEADY STATE DIFFUSION

Those who dislike calculus, and especially partial differential equations can skip this section, but the conclusions should be noted.

In certain circumstances, we are interested in the transient or unsteady state situation which exists when diffusion starts between two volumes which are brought into contact. This is shown conceptually in Figure 7.4, in which a "shutter" is removed, exposing a concentration discontinuity. The two regions proceed to mix and chemical diffuses, eventually achieving homogeneity. Environmentally, this situation is encountered when a volume of fluid (e.g., water) moves to an interface and there contacts another phase (e.g., air) containing a solute with a different fugacity. Volatilization may then occur over a period of time.

There are now three variables, concentration (C), position (y), and time (t). If we consider a volume of $A\Delta y$, as shown in Figure 7.4, then the flux in is $-BA$ dC/dy and the flux out is $-BA(dC/dy + \Delta y d^2C/dy^2)$, while the accumulation is $A\Delta y\Delta C$ in the time increment Δt. It follows that

$$-BA \ dC/dy + BA(dC/dy + \Delta y d^2C/dy^2) = A\Delta y\Delta C/\Delta t$$

or as Δy and Δt tend to zero

$$Bd^2C/d^2y = dC/dt$$

This is Fick's Second Law. Solution of this partial differential equation requires two boundary conditions, usually initial concentrations at specified positions. A particularly useful solution is the "penetration" equation which describes diffusion into a slab of fluid which is brought into contact with another slab of constant concentration. The boundary conditions are

$$C = C_S \text{ at } y = 0 \text{ at all times}$$
$$C = 0 \text{ for } y > 0 \text{ at } t = 0$$

Solution is easiest if some hindsight is invoked to suggest that the dimensionless group X or $(y/\sqrt{4Bt})$ will occur in the solution. Interestingly, this is of the same form as the initial definition of B as $y^2/2t$.

It can be shown that

$$C = C_S[1 - 2\sqrt{\pi}) \int_0^X \exp(-X^2)dX] = C_s \ (1 - \text{erf}(X))$$

where $X = y/\sqrt{4Bt}$

Unfortunately, this integral, which is known as the Gauss Error integral or probability function or error function, cannot be solved analytically; thus, tabulated values must be used. The error function has the property that it is zero when X is zero, and effectively becomes unity when X is 3 or larger. Its value can be found in many tables of mathematical functions. A convenient approximation is

$$\text{erf } (X) = 1 - \exp(-0.746X - 1.101 \ X^2)$$

which is quite accurate when X exceeds 0.75. When X is less than 0.5, erf (X) is approximately $1.1X$. The penetration solution shown in Figure 7.4 illustrates the very rapid initial transfer close to the interface, followed by slower

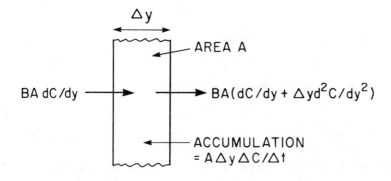

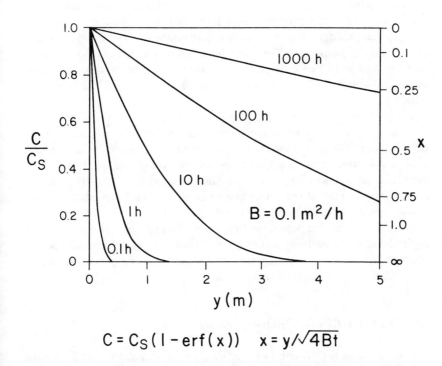

$$C = C_S(1 - erf(x)) \quad x = y/\sqrt{4Bt}$$

Figure 7.4.   The nature of unsteady state or penetration diffusion.

penetration which occurs later as the concentration gradient becomes smaller. Now the transfer rate at the boundary ($y = 0$) can be shown to be

$$B(dC/dy)_{y=0} = C_sA\sqrt{B/\pi t}$$

Over a time t, the total flux (mol) becomes

$$C_sA\sqrt{4Bt/\pi}$$

The average flux is then obtained by dividing by t

$$C_sA\sqrt{4B/\pi}t \text{ mol/h}$$

But, since the average flux is C_sk_M, the average mass transfer coefficient k_M which applies over this time must be $\sqrt{4B/\pi t}$.

The mass transfer coefficient k_M under these transient conditions thus depends on the time of exposure (short exposures giving a large k_M) and on the square root of diffusivity. This contrasts with the steady state solution, in which k_M is proportional to diffusivity, and independent of time. The reason for this behavior is that k_M is apparently very large initially because the concentration gradient is large. It falls in inverse proportion to $\sqrt{t}$; thus, the average also falls in this proportion. The lower dependence on diffusivity (to the power of 0.5 instead of 1.0) arises because not all the transferring mass has to diffuse the total distance; much of it goes into "storage" during the transient concentration buildup.

The problem now arises environmentally—which definition of k_M is correct, $B/\Delta y$ or $\sqrt{4B/\pi t}$? There are several indicators. If the contact time between phases is long, and the amount transferred exceeds the capacity of the phases, it is likely that steady state applies, and we should use $B/\Delta y$. Conversely, if the contact time is short, we can expect to use $\sqrt{4B/\pi t}$. If we measure the transfer rates at several temperatures, and thus different diffusivities, or measure the transfer rate of different chemicals of different B, then plot k_M versus B on log-log paper, the slope of the line will be 1.0 if steady state applies, and 0.5 if unsteady state applies. In practice, an intermediate power of about 2/3 often applies, suggesting that we have mostly penetration diffusion followed by a touch of steady state diffusion.

7.6   DIFFUSION IN POROUS MEDIA

When a solute is diffusing in air or water, its movement is restricted only by collisions with other molecules. If solid particles or phases are also present, the solid surfaces will also block diffusion and slow the net velocity. Environmentally, this is important in sediments in which a solute may have been deposited at some time in the past, and from which it is now diffusing back to the overlying water. It is also important in soils from which pesticides may be volatilizing. It is thus essential to address the question—by how much does the presence of the solid phase retard diffusion? We assume that the solid particles are in contact, but there remains a tortuous path for diffusion, otherwise there is no access route and the diffusivity would be zero.

The process of diffusion is shown schematically in Figure 7.5, in which it is apparent that the solute experiences two difficulties. First, it must take a more tortuous path (which can be defined by a tortuosity factor, F_Y, the ratio of tortuous distance to direct distance). Second, it does not have available the full area for diffusion, i.e., it is forced to move through a smaller area, which can be defined using an area factor, F_A. This area factor, F_A, is equal to the void fraction, i.e., the fraction of the total volume which is fluid, and thus accessible to diffusion. It can be argued that the tortuosity factor, F_Y, is related to void fraction, v, possibly inversely to the power 0.5; thus, in total, we can postulate that the effective diffusivity in the porous medium, B_E, can be related to the molecular diffusivity, B, by

$$B_E = BF_A/F_Y = Bv^{1.5}$$

Such a relationship is found for packings of various types of solids, as discussed by Satterfield (1970). This equation may be seriously in error since (a) the effective diffusivity is sensitive to the shape and size distribution of the particles, (b) there may also be "surface diffusion" along the solid surfaces, and (c) the solute may become trapped in "cul de sacs" or become sorbed on active sites. At least the equation has the correct property that it reduces to intuitively correct limits when v is unity and zero. There is no substitute for actual experimental measurements using the soil or sediment and solute in question.

For soils it is usual to employ the Millington-Quirk (MQ) expression for diffusivity as a function of air and water contents. An example is in the soil diffusion model of Jury et al. (1983).

The MQ expression uses air and water volume fractions v_A and v_W and calculates effective air and water diffusivities as follows

$$B_{AE} = B_A v_A^{10/3}/(v_A + v_W)^2$$
$$B_{WE} = B_W v_W^{10/3}/(v_A + v_W)^2$$

where B_A and B_W are the molecular diffusivities and B_{AE} and B_{WE} the effective diffusivities. Inspection of these equations shows that they reduce to a similar form to that presented earlier. If v_W is zero, B_{AE} is proportional to v_A to the power 1.33 instead of 1.5.

Occasionally there is confusion when selecting the concentration driving force which is to be multiplied by B_E. This should be the concentration in the diffusing medium, not the total concentration including sorbed form. In

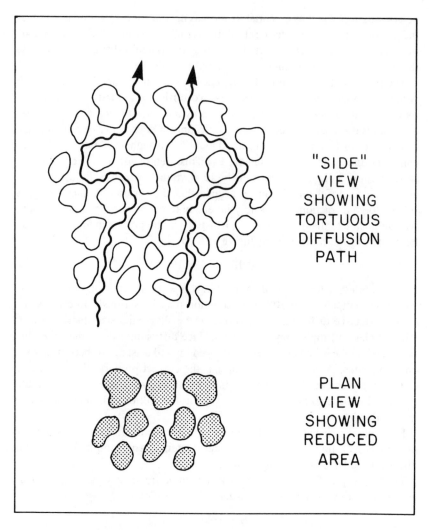

"SIDE" VIEW SHOWING TORTUOUS DIFFUSION PATH

PLAN VIEW SHOWING REDUCED AREA

Figure 7.5.   Diffusion in a porous medium.

sediments the pore water concentration may be 0.01 mol/m³ but the total sorbed plus pure water, i.e., bulk concentration, may be 10 mol/m³. B_E should then be multiplied by 0.01, not 10. In some situations (regrettably) the total concentration (10) is used, in which case B_E must be redefined to be a much smaller "effective diffusivity," i.e., by a factor of 1000.

In sediments it is suspected that much of the chemical present in the pore or interstitial water, and thus available for diffusion, is associated with colloidal organic material. These colloids can diffuse; thus, the diffusing chemical has the option of diffusing in solution or piggybacking on the colloid. From the Stokes-Einstein equation the diffusivity B is approximately inversely proportional to the molecular radius. A typical chemical may have a molecular mass of 200 and a colloid an equivalent molecular mass of 6000, i.e., it is a factor of 30 larger in mass and volume, but only a factor of $\sqrt[3]{30}$ or about 3 in radius. The colloid diffusivity will thus be about one third that of the dissolved molecule. But if 90% of the chemical in the pore water is sorbed to colloids, the colloidal diffusion rate will exceed that of the dissolved form. It is thus necessary to calculate and interpret the contributing diffusion processes, since it may not be obvious which route is faster.

7.7 DIFFUSION BETWEEN PHASES

This discussion has treated only one phase, but in reality we are most interested in situations where the chemical is migrating from one phase to another. It thus encounters two diffusion regimes, one on each side of the interface. Environmentally, such regimes occur principally in air-water exchange, but the same principles apply to diffusion from sediment to water, soil to water and to air, and even to biota-water exchange.

An immediate problem arises at the interface where it is apparent that the chemical will undergo a concentration "jump" from one equilibrium value to another. The chemical may even migrate across the interface from low to high concentration. Clearly, whereas concentration difference was a satisfactory "driving force" for diffusion *within* one phase, it is not satisfactory for describing diffusion *between* two phases. When diffusion is complete, the chemical's fugacities on both sides of the interface will be equal; thus, we can use fugacity as a "driving force" or as a measured "departure from equilibrium." Indeed, fugacity is the fundamental driving force in both cases, but there is no necessity to introduce it for one-phase systems, because only one Z applies and fugacity difference is proportional to concentration difference.

Traditionally, interphase transfer processes have been characterized using the Whitman Two Resistance mass transfer coefficient (MTC) approach (Whitman 1923), in which departure from equilibrium is characterized using a partition coefficient, or in the case of air-water exchange, a Henry's law constant. We derive this approach below for air-water exchange using the Whitman approach and following Liss and Slater (1974), who first applied it to transfer of gases between the atmosphere and the ocean, and Mackay and

Leinonen (1975), who applied the same principles to other organic solutes. We later derive the same equations in fugacity format.

Figure 7.6 illustrates an air-water system in which a solute (chemical) is diffusing at steady state from solution in water at concentration C_W (mol/m^3) to the air at concentration C_A mol/m^3, or at a partial pressure P (Pa), equivalent to $C_A RT$. We assume that the solute is transferred relatively rapidly in the bulk of the water by eddies; thus, the concentration gradient is slight. But as it approaches the interface, the eddies are damped, diffusion slows, and a larger concentration gradient is required to sustain a steady diffusive flux. A mass transfer coefficient, k_W, applies over this region. The solute reaches the interface at a concentration C_{WI}, then abruptly changes to C_{AI}, the air phase value. The question arises as to whether or not there is a significant resistance to transfer at the interface. It appears that if it does exist, it is small and unmeasurable. In any event we do not know how to estimate it, so it is convenient to ignore it and assume that equilibrium applies. We thus proceed on the basis that there is no interfacial resistance, and C_{WI} and C_{AI} are in equilibrium or

$$C_{AI}/C_{WI} = K_{AW} = Z_A/Z_W = H/RT$$
$$\text{and } C_{AI}/K_{AW} = C_{WI}$$

The solute then diffuses in the air from C_{AI} to C_A with a mass transfer coefficient k_A. We can write the flux equations for each phase

$$N = k_W A(C_W - C_{WI}) = k_A A(C_{AI} - C_A) \text{ mol/h}$$

or more conveniently

$$C_W - C_{WI} = N/k_W A$$
$$C_{AI} - C_A = N/k_A A$$
$$\text{or } C_{AI}/K_{AW} - C_A/K_{AW} = N/(k_A A K_{AW})$$
$$\text{thus } C_{WI} - C_A/K_{AW} = N/(k_A A K_{AW})$$

adding the first and last equations to eliminate C_{WI} gives

$$C_W - C_A/K_{AW} = N\{1/k_W A + 1/k_A A K_{AW}\} = N/k_{OW} A$$
$$\text{or } N = k_{OW} A(C_W - C_A/K_{AW}) = k_{OW} A(C_W - P/H)$$
$$\text{where } 1/k_{OW} = 1/k_W + 1/k_A K_{AW} = 1/k_W + RT/Hk_A$$

The term k_{OW} is an "overall" mass transfer coefficient which contains the individual k_W and k_A terms and K_{AW}. The significance of the addition of reciprocal k terms is perhaps best understood by viewing the process in terms of resistances rather than conductivities, where the resistance R, is 1/k in the same sense that the electrical resistance (ohms) is the reciprocal of conductivity (siemens). The overall resistance, R_O, is then given by

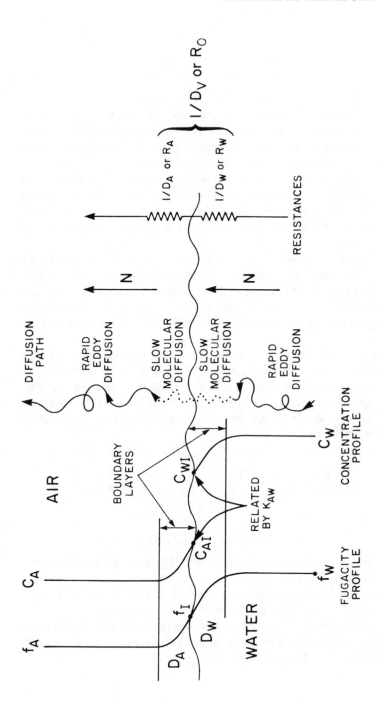

Figure 7.6. Mass transfer between two phases: the two-resistance concept.

$$R_W = 1/(k_W A)$$
$$R_A = RT/(Hk_A A) = 1/(K_{AW} A k_A)$$
$$R_O = R_W + R_A = 1/(k_{OW} A)$$

which is equivalent to the equation for $1/k_{OW}$ above.

Because the resistances are in series, they add, and the total reciprocal conductivity is the sum of the individual reciprocal conductivities. The reason that K_{AW} enters the summation of resistances is that K_{AW} controls the relative values of the concentrations in air and water. If K_{AW} is large, C_{Wi} is small compared to C_{AI}, thus the concentration difference $(C_W - C_{WI})$ will be small compared to $(C_{AI} - C_A)$, and the flux N will be constrained by the value of $k_W(C_W - C_{WI})$. In general, diffusive resistances tend to be largest in phases where the concentrations are lowest, and thus the concentration gradients are lowest.

Typical values of k_A and k_W are, respectively, 10 and 0.1 m/h; thus, the resistances become equal when K_{AW} is 0.01 or H is approximately 25 Pa m³/mol. If H exceeds 250 Pa m³/mol, the air resistance, R_A, is probably less than one-tenth of R_W and may be ignored. Conversely, if H is less than 2.5 Pa m³/mol, the water resistance R_W is less than one-tenth of R_A and it can be ignored.

Interestingly, when H is large, k_W tends to equal k_{OW}, and if C_A or P/H is small, the flux N becomes simply $k_W A C_W$. This group does not contain H; thus, the evaporation rate becomes independent of H or of vapor pressure. At first sight this is puzzling. The reason is that, if H or P^s is high enough, its value ceases to matter because the overall rate is limited only by the diffusion resistance in the water phase.

An overall mass transfer coefficient k_{OA} can also be defined as

$$1/k_{OA} = 1/k_A + H/RTk_W = 1/k_A + K_{AW}/k_W$$

and $$N = k_{OA} A(C_W K_{AW} - C_A) = k_{OA} A(C_W H - P)/RT$$

It follows that

$$k_{OW} = k_{OA} K_{AW} = k_{OA} H/RT$$

If H is low, k_{OA} approaches k_A, and when P/H is small, the flux approaches $k_A A C_W K_{AW}$ or $k_A A C_W H/RT$. In such cases, volatilization becomes proportional to H and may be negligible if H is very small. In the limit, when H is zero (as with sodium chloride) volatilization does not occur at all.

Figure 7.7 is a plot of the saturation concentrations in air and water, i.e. $\log P^s/RT$ versus $\log C^s$ on which the location of certain solutes is indicated. Recalling that, since P^s equals HC^s, K_{AW} is H/RT

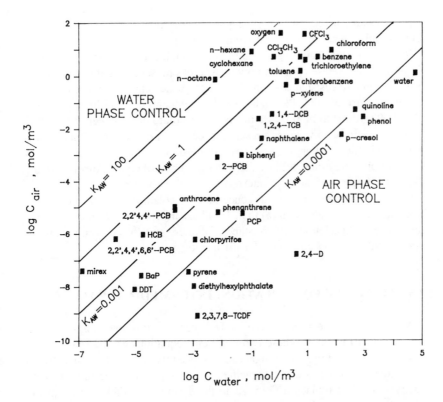

Figure 7.7. Plot of log air solubility or P^S/RT versus log water solubility for selected chemicals. Note that chemicals of equal H or K_{AW} or P^S/C^S lie on the same 45° diagonal.

$$\log P^S = \log H + \log C^S$$
$$\text{and } \log P^S/RT = \log K_{AW} + \log C^s$$

thus compounds of equal H or K_{AW} will lie on the same 45° diagonal. Compounds of H > 250 Pa.m³/mol or $K_{AW} > 0.1$ lie to the upper left, are volatile, and are water phase diffusion controlled. Those of H < 2.5 or $K_{AW} < 0.001$ lie to the lower right, are relatively involatile, and are air phase diffusion controlled. There is an intermediate band in which both resistances are important.

It is interesting to note that a homologous series of chemicals tends to lie along a 45° diagonal of constant H or K_{AW}. Substituting methyl groups or chlorines for hydrogen tends to reduce both vapor pressure and solubility by a factor of 4 to 6; thus, H tends to remain relatively constant and the series retains a similar ratio of air and water resistances. Paradoxically, reducing vapor pressure as one ascends such a series does not reduce evaporation rate from solution since it is H that controls, not P^S.

It is noteworthy that oxygen and most low molecular weight hydrocarbons lie in the water phase resistant region, whereas most oxygenated organics lie in the air phase resistant region. The H for water can be deduced from its vapor pressure of 2000 Pa at 20°C and its concentration in the water phase of 55000 mol/m³ to be 0.04 Pa m³/mol. If a solute has a lower H than this, it may *concentrate* in water as a result of faster water evaporation, but of course humidity in the air alters the water evaporation rate. Water evaporation is entirely air phase resistant because the water need not, of course, diffuse through the water phase to reach the interface. It is already there.

Certain inferences can be made concerning the volatilization rate of one solute from another, provided that (a) their H values are comparable, i.e., the same resistance or distribution of resistances applies, and (b) corrections are applied for differences in molecular diffusivity.

7.8   FUGACITY FORMULATION: INTERMEDIA D VALUES

As was discussed earlier, we can use D values instead of mass transfer coefficients and diffusivities. These two-resistance equations can be reformulated in fugacity terms to yield an algebraic result equivalent to the concentration version. If the water and air fugacities are f_W and f_A and the interfacial fugacity is f_I, then replacing C by Zf in the steady state Fick's law equation yields

$$N = k_W A(C_W - C_{WI}) = k_W A Z_W(f_W - f_I) = D_W(f_W - f_I) \text{ mol/h}$$
$$\text{and } N = k_A A(C_{AI} - C_A) = k_A A Z_A(f_I - f_A) = D_A(f_I - f_A) \text{ mol/h}$$

where $D_W = k_W A Z_W$ and $D_A = k_A A Z_A$. This is illustrated in Figure 7.6.

Now the interfacial fugacity f_I is not known or measurable, thus it is convenient to eliminate it by adding the equations in rearranged form, namely:

$$f_W - f_I = N/D_W$$
$$\text{and } f_I - f_A = N/D_A$$
$$\text{adding gives } (f_W - f_A) = N(1/D_W + 1/D_A) = N/D_V$$
$$\text{and } N = D_V(f_W - f_A)$$
$$\text{where } 1/D_V = 1/D_W + 1/D_A = 1/k_W A Z_W + 1/k_A A Z_A$$

The groups $1/D_A$ and $1/D_W$ are effectively resistances which add to give the total resistance $1/D_V$. It can be shown that

$$D_V = k_{OW} A Z_W = k_{OA} A Z_A$$

thus $k_{OW}/k_{OA} = Z_A/Z_w = K_{AW}$ as before

The net volatilization rate, $D_V(f_W - f_A)$ can be viewed as the algebraic sum of an upward volatilization rate, $D_V f_W$, and a downward absorption rate $D_V f_A$.

7.9 MEASURING TRANSPORT D VALUES

Measuring nondiffusive D values is, in principle, simply a matter of measuring Z and G, the latter usually being the problem. Flows of air, water, particulate matter, rain, and food can be estimated directly. The more difficult situations involve estimation of the rates of deposition of aerosols and sedimenting particles in the water column. The obvious approach is to place a bucket, tray, or a sticky surface in the depositing region and measure the amount collected. This method can be criticized in that the presence of the bucket alters the hydrodynamic regime, and thus the settling rate. This problem is acute when estimating aerosol deposition rates on foliage in a field or forest where the boundary layer is highly disturbed. Measurements of resuspension rates are particularly difficult because the resuspension event may be triggered periodically by a storm or flood, or by an especially energetic fish chasing prey at the bottom. Regrettably, the sediment-water interface is not easily accessible; thus, measurements are few, difficult, and expensive.

Measurement of mass transfer coefficients or diffusive D values usually involves setting up a system in which there is a known area and fugacity driving force $(f_1 - f_2)$ and the capacity to measure N, leaving the overall transport D value (or the mass transfer coefficient) as the only unknown in the flux equation.

Air phase mass transfer coefficients (MTCs) can be estimated by measuring the evaporation rate of a pool of pure liquid, or even the sublimation rate of a volatile solid. The interfacial partial pressure, fugacity, or concentration of the solute can be found from vapor pressure tables. The concentration some distance from the surface can be zero if adequate air circulation is arranged, thus ΔC or Δf is known. The pool can be weighed periodically to determine N, and area A can be measured directly; thus, the MTC or evaporation D value is the only unknown.

Worked Example 7.3

A tray (50 cm by 30 cm in area) contains benzene at 25°C (vapor pressure 12700 Pa). The benzene is observed to evaporate into a brisk air stream at a rate of 585 g/h. What are D and k_M?

Since the molecular mass is 78 g/mol, N is 585/78 or 7.5 mol/h.

$$\Delta f \text{ is } (12700 - 0) \text{ Pa}$$
$$D = 7.5/12700 = 5.9 \times 10^{-4} \text{ mol/Pa.h.}$$

A is 0.5×0.3 or 0.15 m^2. Z_A is $1/RT$ or 4.04×10^{-4}. Since D is $k_M A Z_A$, k_M is 9.7 m/h.

In conventional units ΔC is $12700/RT$ or 5.13 mol/m^3.

$$N = k_M A \Delta C$$

thus $k_M = 9.74$ as before.

Obviously the two approaches are algebraically equivalent. By using an experimental system of this type the dependence of k_M on wind speed can be measured and correlated.

Measurement of overall intermedia D values or MTCs is similar in principle, Δf applying between two bulk phases. A convenient method of measuring water to air transfer is to dissolve the solute in a tank of water, blow air across the surface to simulate wind, and measure the evaporation rate indirectly by following the decrease in concentration in the water with time. If the water volume is V m^3, area is A m^2 and depth is Y m then

$$N = VdC_W/dt = -k_{OW}A(C_W - C_A/K_{AW})$$

where C_A and C_W are concentrations in air and water, and k_{OW} is the overall MTC. Assuming C_A to be zero, and integrating from an initial condition C_{WO} and f_{WO}

$$C_W = C_{WO} \exp(-k_{OW}At/V) = C_{WO} \exp(-k_{OW}t/Y)$$
$$\text{or } f_W = f_{WO} \exp(-D_V t/VZ_W)$$

Plotting C_W on semilog paper vs linear time gives a measurable slope of $-k_{OW}/Y$, from which k_{OW} can be estimated. A system of this type has been described by Mackay and Yuen (1983) and is illustrated in Figure 7.8.

A very useful quantity is the evaporation half-life, which is $0.693Y/k_{OW}$. Often an order of magnitude estimate of this time is sufficient to show that volatilization is unimportant, or it dominates other processes such as reaction.

Measurement of the individual contributing air and water D values or MTCs is impossible because the interfacial concentrations cannot be measured. But if the evaporation rates of a series of chemicals of different K_{AW} are measured it is possible to deduce k_W and k_A or D_W and D_A.

The relationship $1/k_{OW} = 1/k_W + 1/K_{AW}k_A$ suggests plotting, as in Figure 7.8, $1/k_{OW}$ versus $1/K_{AW}$ for a series of chemicals. The intercept will be $1/k_W$ and the slope $1/k_A$.

A correction may be necessary for molecular diffusivity differences. Essentially, k_W is being measured by selecting chemicals of high K_{AW} for which the term $1/k_A K_{AW}$ is negligible. Alkanes, oxygen, or inert gases are

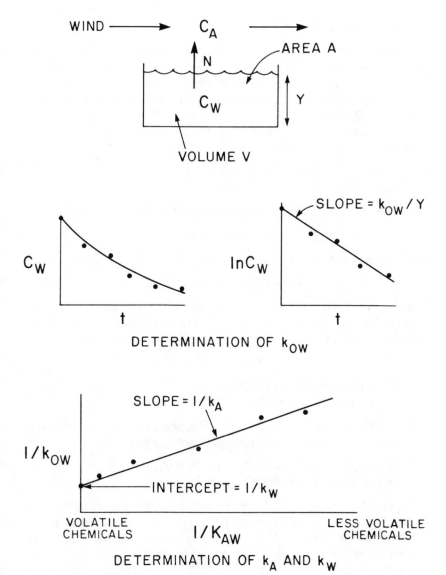

Figure 7.8.   Measurement of intermedia D values and mass transfer coefficients.

convenient. k_A is measured by choosing chemicals of low K_{AW} so that $1/k_A K_{AW}$ is large compared to $1/k_W$. Alcohols are convenient for this purpose.

Worked Example 7.4

A tank contains 2 m³ of water at 25°C, 50 cm deep with dissolved benzene ($K_{AW} = 0.22$) and naphthalene ($K_{AW} = 0.017$), each at a concentration of 0.1 mol/m³. After 2 hours these concentrations have dropped to 47.1 and 63.9% of their initial value, respectively. What are the overall and individual MTCs and D values?

In each case $C_W = C_{WO} \exp (-k_{OW}t/Y)$, Y being 0.5 m. Thus $k_{OW} = - (Y/t) \ln (C_W/C_{WO})$. Substituting gives

$$Benzene \quad k_{OW} = 0.188 \text{ m/h}$$
$$Naphthalene \quad k_{OW} = 0.112 \text{ m/h}$$

Assuming each k_{OW} to be made up of identical k_W and k_A values, i.e.,

$$1/k_{OW} = 1/k_W + 1/K_{AW}k_A$$

two equations can be written and solved for k_W and k_A giving

$$k_W = 0.20 \quad k_A = 15 \text{ m/h}.$$

In fugacity format, Z_A is 4.04×10^{-4} (both), Z_W is 1.836×10^{-3} (benzene) and 23.7×10^{-3} (napthalene); thus, the fugacities are initially 54 and 4.22 Pa, falling to 25 and 2.72 Pa. The D_V values are obtained from

$$f_W = f_{WO} \exp(-D_V t/VZ_W)$$

D_V for benzene = 1.38×10^{-3}
D_V for naphthalene = 10.6×10^{-3}

Now $1/D_V = 1/D_A + 1/D_W$, the D_A value being common to both chemicals, only D_W containing the variable Z_W. Solving gives

Benzene $D_A = 0.0242$ $D_W = 0.00146$
Naphthalene $D_A = 0.0242$ $D_W = 0.0190$

In practice it is unwise to rely on only two chemicals, it being better to use at least five covering a wide range of K_{AW} values. The air phase resistance when viewed as $1/D_A$ is 41.3 units in both cases, but the water phase resistance for benzene is 685, while for naphthalene it is 52.

Example 7.5

10 kg each of benzene, 1,4-dichlorobenzene, and p-cresol are spilled into a pond 5 m deep with an area of 1 km². If k_W is 0.1 m/h and k_A is 10 m/h what will be the times necessary for half of each chemical to be evaporated? Use the property data from Chapter 3, and ignore other loss processes.

Oxygen Transfer

Some of the earliest environmental modeling was of oxygen transfer to oxygen depleted rivers in which a "reaeration constant," k_2, was introduced (with units of reciprocal time) using the equation

$$dC_W/dt = k_2(C_E - C_W) \text{ mol/m}^3\text{h}$$

where C_E is the equilibrium solubility of oxygen in water. Another term is usually included for oxygen consumption, but we ignore it here. Now, if the volume in question is one square meter in horizontal area and Y m deep, it will have a volume of Y m^3 and the flux N will be YdC_w/dt mol/h. But

$$N = k_M(C_E - C_W) = YdC_W/dt = Yk_2(C_E - C_W)$$

k_M is thus equivalent to Yk_2. A typical k_2 of 1 day^{-1} in a river of depth 2.4 m corresponds to a mass transfer coefficient of 2.4 m/day or 0.1 m/h. Oxygen reaeration rates can thus be used to estimate mass transfer coefficients for other solutes of similar (large) H such as alkanes. Indeed, an ingenious experimental approach for determining k_2 for oxygen is to use a volatile hydrocarbon, such as propane, as a tracer, thus avoiding the complications of biotic oxygen consumption or generation which confound environmental measurements of oxygen concentration change. It is erroneous to use k_2 to estimate the rate of volatilization of a chemical of low H since k_2 contains negligible air phase resistance information. A correction should also be applied for the effect of molecular diffusivity, preferably using the dimensionless form of diffusivity, the Schmidt Number raised to a power such as 0.5 or 0.67.

Other Systems

This technique of probing interfacial MTCs by measuring N for various chemicals can be applied in other areas. When chemical is taken up by fish it appears that it passes through one or more water layers, and one or more organic membranes in series. By analogy with air-water transfer we can write an organic-water transfer equation simply by replacing A by O giving

$$N = k_{OW}A(C_W - C_O/K_{OW})$$

where

$$1/k_{OW} = 1/k_W + 1/k_OK_{OW}$$

or more conveniently changing to an overall organic phase MTC k_{OO}

$$N = k_{OO}A(C_WK_{OW} - C_O)$$

where

$$1/k_{OO} = 1/k_O + K_{OW}/k_W$$

A plot of $1/k_{OO}$ versus K_{OW}, the organic-water or octanol water partition coefficient, gives $1/k_O$ as intercept and $1/k_W$ as slope. This is essentially the fish bioconcentration equation in disguise, which is conventionally written

$$dC_F/dt = k_1 C_W - k_2 C_F$$

If V_F is fish volume and C_O is C_F/L, where L is the volume fraction lipid (equivalent to octanol) in the fish, it follows that

$$dC_F/dt = (k_{OO}A/V_F)(C_W K_{OW} - C_F/L)$$

k_1 is obviously $K_{OW}k_{OO}A/V_F$ and k_2 is $(k_{OO}A/(V_F L))$. Then k_1/k_2 is LK_{OW}, the bioconcentration factor. The area of the respiring gill surface is uncertain, as is k_{OO}, so it is convenient to lump these uncertainties in one unknown k_2. This suggests plotting $1/k_2$ versus K_{OW} to obtain quantities containing k_O and k_W as well as A, V_O, and L. Such a plot was compiled by Mackay and Hughes (1984), yielding estimates of $V_O L/Ak_O$ and $V_O L/Ak_W$ which have dimensions of time.

Another example is the penetration of chemicals through the waxy cuticles of leaves in which there are air and wax resistances in series. Kerler and Schonherr (1988) have measured such penetration rates for a variety of chemicals, and Schramm et al. (1987) have attempted to model chemical uptake by trees, using this two-resistance approach. A plant's principal problem in life is to manage its water budget and avoid at all costs excessive loss of water through leaves. It accomplishes this by forming a waxy layer through which water has only a very slow diffusion rate. Diffusivities are phenomenally low, leading to very low MTCs and D values for water. The plant thus exploits this two-resistance approach to conserve water. If only governments could manage their budgets with the same efficiency!

7.10   COMBINING SERIES AND PARALLEL D VALUES

Having introduced these transport D values and shown how they combine when describing resistances in series, it is useful to set out the general flux equation for any combination of transport processes in series or parallel.

Each transport process can be quantified by a D value (deduced as GZ, kAZ, or BAZ/Y) and applied between two points in space such as a bulk phase and an interface, or between two bulk phases. It is helpful to prepare an arrow diagram of the processes showing the connections, as illustrated in Figure 7.9. Diffusive processes are reversible, so they actually consist of two arrows in opposing directions with the same D value, but driven by different source fugacities.

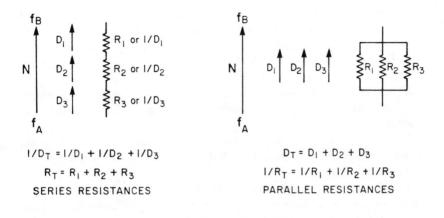

Figure 7.9.   Series and parallel combinations of D values.

When processes apply in parallel between common points, the D values add. An example is wet and dry deposition from bulk air to bulk water.

$$D_{TOTAL} = D_1 + D_2 + D_3 \text{ etc.}$$

When processes apply in series, the resistances add, or correspondingly, the reciprocal D values add to give a reciprocal total

$$1/D_{TOTAL} = 1/D_1 + 1/D_2 + 1/D_3 \text{ etc.}$$

An example is the addition of air and water boundary layer resistances, as discussed earlier.

It is possible to assemble numerous combinations of series and parallel processes linking bulk phases and interfaces. These situations can be viewed as electrical analogs with voltage being equivalent to fugacity, resistance equivalent to 1/D and current equivalent to flux (mol/h). Figure 7.9 gives some examples.

In air-water exchange there can be deposition by (a) dry particle deposition, (b) wet particle deposition, (c) rain dissolution, and diffusive absorption-volatilization.

The soil-air exchange example involves parallel diffusive transport from bulk soil to the interface in water and air, followed by a series air boundary layer diffusion step.

The sediment-water example is similar, having parallel diffusive paths for chemical transport in water and in association with organic colloids. The principal difficulty is estimation of the diffusivity of the colloids.

Even more complex combinations can be compiled for transport processes into and within organisms, this being essentially the science of pharmacokinetics. An example is an attempt to express transport of styrene in humans (Paterson and Mackay, 1986), which is discussed later in Chapter 8.

7.11 LEVEL III CALCULATIONS

In this chapter we have examined the nature of molecular and eddy diffusivities, introduced the concept of mass transfer coefficients (k), and treated the problem of two resistances occurring in series as material diffuses from one phase to another. Two new D values have been introduced, a kAZ product and a BAZ/Δy product. We can treat situations in which various D values apply in series and in parallel.

In some situations, diffusive D values may be assisted or countered by advective transfer D values. For example, PCB may be evaporating from a water surface into the atmosphere, only to return by association with aerosol particles that fall by wet or dry deposition. We can add D values when the fugacities with which they are multiplied are identical, i.e., the source is the same phase. This is convenient because it makes the equations algebraically simple and enables us to compare the rates at which materials move by various mechanisms between phases.

We thus have at our disposal an impressive set of tools for calculating transport rates between phases. We need Z values, mass transfer coefficients, diffusivities, path lengths, and advective flow rates. Quite complicated models can be assembled describing transfer of a chemical between several media by a number of routes. In general, the total D value for movement from phase A to phase B will not be the same as that from B to A. The reason is that there is often an advective process moving in only one direction. Diffusive processes always have identical D values applying in each direction. We are now able to use these ideas to perform a Level III calculation.

These calculations were suggested and illustrated in a series of papers on fugacity models (Mackay 1979, Mackay and Paterson 1981, 1982, 1991, and Mackay et al. 1985). It is important to emphasize that these models will give the same results as other concentration-based models such as GEOTOX, EXAMS, TOXIWASP and WASTOX, provided that the intermedia transport expressions are ultimately equivalent. A major advantage of the fugacity approach is that an enormous amount of detail can be contained in one D value, which can be readily compared with other D values for quite different processes. It is difficult, on the other hand, to compare a reaction rate constant, a mass transfer coefficient, and a sedimentation rate and identify their relative importance.

Figure 7.10 depicts a simple four-compartment evaluative environment with the intermedia transport processes indicated by arrows. In addition to the reaction and advection D values which were introduced in Level II there are seven intermedia D values. The emission rates of chemicals must now be specified on a medium by medium basis, whereas in Level II only the total emission rate was needed.

Estimating Intermedia D Values

The seven intermedia D values can be estimated from a knowledge of the nature of each contributing process and a variety of flow rates, areas, mass transfer coefficients, etc. It is convenient to subscript air, water, soil, and sediment, respectively, 1, 2, 3, and 4.

The air to water D value D_{12} consists of diffusive absorption (D_V) and nondiffusive wet and dry aerosol deposition. Each D value can be estimated and summed to give D_{12} as discussed later in Chapter 8.2.

The water to air D value D_{21} is D_V for diffusive volatilization and is, of course, the same D_V as for absorption.

The air to soil D value D_{13} is similar to D_{12} but the areas differ and the absorption-volatilization D value is also different.

The soil to air D value D_{31} is for volatilization as discussed in Chapter 8.3.

The water to sediment D value D_{24} represents diffusive transfer plus nondiffusive sediment deposition as discussed in Chapter 8.4.

The sediment to water D value D_{42} represents diffusive transfer plus nondiffusive resuspension. It may also contain transfer of colloids from sediment to water.

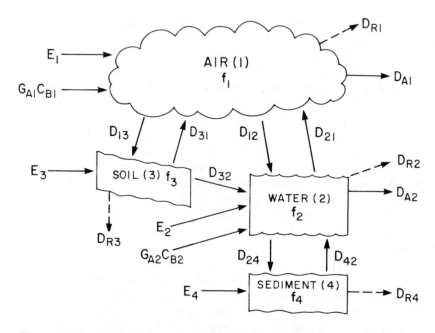

Figure 7.10.   Four compartment Level III diagram.

Finally, the soil to water D value D_{32} consists of nondiffusive water and particle runoff.

There is no water to soil transfer, nor is there sediment-air exchange.

The obvious approach is to gather the required transport data, estimate each D value, then enter them into the mass balance equations and solve for the fugacities, as is described shortly. The modeler has usually no intuitive feeling for the magnitude of the D values; thus, it is not clear if 10^6 mol/h.Pa is a very fast or slow process. It may be useful, for illustrative purposes, to insert a large (fast) or small (slow) D value, but there is no obvious method of selecting appropriate, order-of-magnitude values. This can be done using the relationship between the process D value and the half-time or characteristic time by which the process in question can deplete the source medium of chemical. For example, the volatilization half-life from water to air may be 100 hours. Note that the receiving phase may experience a different half-life, even if the same D value applies, because the VZ group differs. If only this process applies then

$$VZdf/dt = -Df \quad \text{and} \quad f = f_0 \exp(-Dt/VZ)$$

from which the half-life is clearly 0.693 VZ/D. The V and Z apply to the source phase and D is the source-to-destination D value. If a half-life $t_{1/2}$ can be suggested, then D will be $0.693\ VZ/t_{1/2}$. Short half-lives represent large D values and fast, important processes. It is useful to calculate a half-life or a characteristic time VZ/D to ensure that it is reasonable.

Level III Mass Balance Equations and Solution

We now write the mass balance equations for each medium as follows.

Air (subscript 1)
$$E_1 + G_{A1}.C_{B1} + f_2D_{21} + f_3D_{31} = f_1(D_{12} + D_{13} + D_{R1} + D_{A1}) = f_1D_{T1}$$
Water (subscript 2)
$$E_2 + G_{A2}C_{B2} + f_1D_{12} + f_3D_{32} + f_4D_{42} =$$
$$f_2(D_{21} + D_{24} + D_{R2} + D_{A2}) = f_2D_{T2}$$
Soil (subscript 3)
$$E_3 + f_1D_{13} = f_3(D_{31} + D_{32} + D_{R3} + D_{A3}) = f_3D_{T3}$$
Sediment (subscript 4)
$$E_4 + f_2D_{24} = f_4(D_{42} + D_{R4} + D_{A4}) = f_4D_{T4}$$

In each case E_i is the emission rate (mol/h), G_A is the advective inflow rate (m^3/h), C_{Bi} is the advective inflow concentration (mol/m^3), D_{Ri} is the reaction rate D value, and D_{Ai} is the advection rate D value. D_{Ti} is the sum of all loss D values from medium i. Sediment burial, air to stratospheric transfer, and leaching from soil to groundwater can be included as advection processes or as pseudo-reactions. Normally there is no advective loss from soil and sediment; thus, D_{A3} and D_{A4} are zero, but they are included above for completeness.

These four equations contain four unknowns (the fugacities); thus, solution is possible. After some algebra it can be shown that

$$f_2 = (I_2 + J_1J_4/J_3 + I_3D_{32}/D_{T3} + I_4D_{42}/D_{T4})/(D_{T2} - J_2J_4/J_3 - D_{24}.D_{42}/D_{T4})$$
$$f_1 = (J_1 + f_2J_2)/J_3$$
$$f_3 = (I_3 + f_1D_{13})/D_{T3}$$
$$f_4 = (I_4 + f_2D_{24})/D_{T4}$$
where
$$J_1 = I_1/D_{T1} + I_3D_{31}/(D_{T3}.D_{T1})$$
$$J_2 = D_{21}/D_{T1}$$
$$J_3 = 1 - D_{31}D_{13}/(D_{T1}.D_{T3})$$
$$J_4 = D_{12} + D_{32}.D_{13}/D_{T3}$$
and
$$I_i = E_i + G_{Ai}C_{Bi},$$ i.e., the total of emission and advection inputs into each phase.

Unlike the Level II calculation, it is now necessary to specify the emissions into each compartment separately.

Having obtained the fugacities, all process rates can be deduced as Df and a steady state mass balance should emerge in which the total inputs to each medium equals the outputs. The amounts and concentrations can be calculated.

An overall residence time can be calculated as the sum of the amounts present divided by the total input (or output) rate. A reaction residence time can be similarly calculated as the amount divided by the total reaction rate, and a corresponding advection residence time can also be deduced.

Figure 7.11 is an illustrative output from a Level III calculation. This is a comprehensive multimedia picture of chemical emission, advection, reaction, intermedia transport and residence time or persistence. The important processes are now clear and it is possible to focus on them, seeking more accurate rate data. Figure 7.11 contains information about 21 processes, some of which, such as air-water transfer, consist of several contributing processes. The human mind is incapable of making sense of the vast quantity of physical chemical and environmental data which combine to give this picture, without the aid of a tool such as a Level III program.

Mackay et al. (1985, 1991) examined the behavior of a series of chemicals using a model of this type, illustrating their quite different behavior characteristics.

It is possible to extend the model to include more compartments by writing additional mass balance equations. The algebra becomes more tedious and it is probably best to solve the equations by matrix algebra using a standard computer subroutine.

Computer Model

Computer programs are provided on the diskette which undertake the calculation of the multimedia fate of a specified chemical. The user must provide physical chemical (partitioning) properties, reaction half-lives and intermedia transport D values, either directly, or in the form of half-life transfer times. The program can be easily modified to include subroutines to calculate these D values. Assembly of a Level III model for a chemical is a fairly demanding task since there are numerous areas, flows, mass transfer coefficients, and diffusivities to be estimated. To assist in this task, Table 7.1 summarizes equations for intermedia transport D values and Table 7.2 gives suggested, order of magnitude, values for the various parameters which can serve as a starting point for the calculations.

The user is encouraged to calculate these D values, either by hand or by computer; check that the half-times seem reasonable; then conduct a Level III calculation for chemicals of interest, or those specified in Chapter 3. It

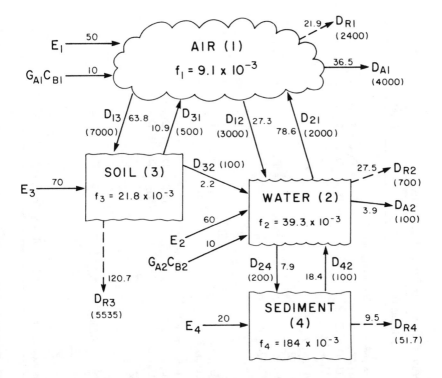

Figure 7.11.   Illustrative Level III output.

is instructive to prepare a mass balance diagram, check that the balance is correct (i.e., input equals output for each compartment), and identify the primary processes which control environmental fate. It may then be appropriate to examine these processes in more detail, seeking more accurate parameter values. Usually the chemical's fate is controlled by relatively few key processes, but it is not always obvious which these are until a Level III calculation is completed.

It is also instructive to undertake a sensitivity analysis by systematically varying input quantities and examining the result in terms of concentrations. Sensitive input parameters can then be identified.

If the parameter values in Table 7.2 are substituted into the equations in Table 7.1, the seven intermedia D values can be expressed as functions of the areas of the media, various Z values, and a series of constants which are essentially combinations of transport quantities such as mass transfer coefficients, flowrates and diffusivities, all with dimensions of velocity (m/h). These constants describe the entire intermedia transport characteristics of all chemicals

Table 7.1. Intermedia Transfer D Value Equations.

Compartment	Process	Individual D	Total D
air(1) - water(2) area A_{12}	diffusion	$D_{VW} = 1/(1/k_{AW} \cdot A_{12} \cdot Z_A + 1/k_{WW} \cdot A_{12} \cdot Z_W)$	
	rain	$D_{RW} = A_{12} \cdot U_R \cdot Z_W$	$D_{12} = D_{VW} + D_{RW} + D_{QW} + D_{DW}$
	wet deposition	$D_{QW} = A_{12} \cdot U_R \cdot Q \cdot v_Q \cdot Z_Q$	$D_{21} = D_{VW}$
	dry deposition	$D_{DW} = A_{12} \cdot U_P \cdot v_Q \cdot Z_Q$	
air(1) - soil(3) area A_{13}	diffusion	$D_{VS} = 1/(1/k_{SA}A_{13}Z_A + Y_3/(A_{13}(B_{AE}Z_A + B_{WE}Z_W)))$	
	rain	$D_{RS} = A_{13} \cdot U_R \cdot Z_W$	$D_{13} = D_{VS} + D_{RS} + D_{QS} + D_{DS}$
	wet deposition	$D_{QS} = A_{13} \cdot U_R \cdot Q \cdot v_Q \cdot Z_Q$	$D_{31} = D_{VS}$
	dry deposition	$D_{DS} = A_{13} \cdot U_P \cdot v_Q \cdot Z_Q$	
soil(3) - water(2) area A_{13}	soil runoff	$D_{SW} = A_{13} \cdot U_{SW} \cdot Z_S$	$D_{32} = D_{SW} + D_{WW}$
	water runoff	$D_{WW} = A_{13} \cdot U_{WW} \cdot Z_W$	$D_{23} = 0$
sediment(4) - water(2) area A_{24}	diffusion	$D_{TX} = 1/(1/k_{XW}A_{24}Z_W + Y_4/B_{WX}A_{24}Z_W)$	
	deposition	$D_{DX} = A_{24} \cdot U_{DX} \cdot Z_P$	$D_{24} = D_{TX} + D_{DX}$
	resuspension	$D_{RX} = A_{24} \cdot U_{RX} \cdot Z_X$	$D_{42} = D_{TX} + D_{RX}$
reaction	bulk phase	$D_{Ri} = k_{Ri} V_i Z_i$	
advection		$D_{Ai} = G_i Z_i$	
transfer to stratosphere		$D_{ST} = U_{ST}(A_{12} + A_{13})Z_A$	
sediment burial		$D_{BX} = U_{BX} \cdot A_{24} \cdot Z_x$	
leaching soil-groundwater		$D_{LS} = U_{LS} \cdot A_{13} \cdot Z_W$	

B_{AE}, B_{WE} and B_{WX} are effective diffusivities deduced from molecular diffusivities and void fractions.

Table 7.2. Order of Magnitude Values of Transport Parameters and Volume Fractions Which Are Used to Convert Molecular Diffusivities to Effective Diffusivities as Described in Chapter 7.6.

Parameter	Symbol	Value
Air side MTC over water	k_{AW}	3 m/h
Water side MTC	k_{WW}	.03 m/h
Transfer rate to stratosphere	U_{ST}	.01 m/h (90 m/y)
Rain rate (m^3rain/m^2area.h)	U_R	9.7×10^{-5} m/h (.85m/y)
Scavenging ratio	Q	200,000
Dry deposition velocity	U_P	10.8 m/h (.003 m/s)
Air side MTC over soil	k_{SA}	1 m/h
Diffusion path length in soil	Y_3	.05 m
Molecular diffusivity in air	B_{MA}	.04 m^2/h
Molecular diffusivity in water	B_{MW}	4.0×10^{-6} m^2/h
Water runoff rate from soil	U_{WW}	3.9×10^{-5} m/h (.34 m/y)
Solids runoff rate from soil	U_{SW}	2.3×10^{-8} m^3/m^2.h (0.0002 m/y)
Water side MTC over sediment	k_{XW}	.01 m/h
Diffusion path length in sediment	Y_4	.005 m
Sediment deposition rate	U_{DX}	4.6×10^{-8} m^3/m^2.h(.0004 m/y)
Sediment resuspension rate	U_{RX}	1.1×10^{-8} m^3/m^2.h(.0001 m/y)
Sediment burial rate	U_{BX}	3.4×10^{-8} m^3/m^2.h(.0003 m/y)
Leaching rate from soil to ground water	U_{LS}	3.9×10^{-5} m^3/m^2.h(.34 m/y)
Volume fraction aerosols	v_Q	2×10^{-11}
Volume fraction air in soil	v_A	0.2
Volume fraction water in soil	v_W	0.3
Volume fraction water in sediment	v_X	0.63

in the system. To a first approximation, the parameter values should be fairly constant for all chemicals because the chemical-to-chemical differences are expressed by the Z values. If a large quantity of data were available for concentrations and inputs of many chemicals in a defined environment, these constants could be "fitted" to the data. This is not presently possible, but it may be in the future. The "fitting" will be assisted by prior knowledge of the likely magnitude of the quantities. A version of the Level III computer program which uses these constants to define the D values is also provided.

Figure 7.12 is an illustrative output of a Level III computer program.

FUGACITY LEVEL III PROGRAM
D value input directly (Level 3B)

Properties of HYPOTHENE

Temperature deg C	27.5
Molecular mass g/mol	200
Vapor pressure Pa	1
Solubility g/m³	20
Solubility mol/m³	.1
Henry's law constant Pa.m³/mol	10
Log octanol-water p-coefficient	4
Octanol-water partn-coefficient	10000
Organic C-water ptn-coefficient	4100
Air-water partition coefficient	4.000 E-03
Soil-water partition coefficient	123
Sedt-water partition coefficient	246
Total emission rate mol/h	200
Total of VZ products	4170480
Total amount moles	156361.4

Phase properties, compositions and rates

Phase	Air	Water	Soil	Sediment
Volume m³	6E+09	7000000	45000	21000
Density kg/m³	1.175556	1000	1500	1500
Frn org carb	0	0	.02	.04
Z mol/m³.Pa	4.000 E-04	.1	12.3	24.6
Adv. flow m³/h	1E+07	1000	0	0
Adv. restime h	600	7000	0	0
Conc in mol/m³	.000001	.01	0	0
Inflow mol/h	10	10	0	0
Emission mol/h	50	60	70	20
Rct halflife h	693	693	69.3	6930

(continued on next page)

(continued from previous page)

VZ mol/Pa	2400380	700000	553500	516600
Fugacity Pa	9.115 E-03	3.929 E-02	.0218	.1836
Conc mol/m³	3.646 E-06	3.929 E-03	.2682	4.519
Conc g/m³	7.293 E-04	.7859	53.65	903.8
Conc μg/g	.6204	.7859	35.76	602.5
Amount mol	21880.71	27509.43	12072.22	94899.06
Amount %	13.99	17.59	7.720	60.69
D rct mol/Pa.h	2400.38	700	5535	51.660
D adv mol/Pa.h	4000.634	100	0	0
Rct rate mol/h	21.88	27.50	120.7	9.489
Adv rate mol/h	36.46	3.929	0	0
Reaction %	9.945	12.50	54.87	4.313
Advection %	16.57	1.786	0	0

Intermedia Data.   Half times (h) Eq. flows (m³/8) D values Rates (mol/h)

Air to water	5.5449E+02	7.4988E+03	3.0000E+03	2.7347E+01
Air to soil	2.3764E+02	1.7497E+07	7.0000E+03	6.3809E+01
Water to air	2.4255E+02	2.0000E+04	2.0000E+03	7.8598E+01
Water to sediment	2.4255E+03	2.0000E+03	2.0000E+02	7.8598E+00
Soil to air	7.6715E+02	4.0650E+01	5.0000E+02	1.0905E+01
Soil to water	3.8358E+03	8.1301E+00	1.0000E+02	2.1811E+00
Sediment to water	3.5800E+03	4.0650E+00	1.0000E+02	1.8370E+01

Total chemical input	mol/h	220
Total chemical output	mol/h	220
Output by reaction	mol/h	179.6022
Output by advection	mol/h	40.39776
Total output	mol/h	220
Overall residence time	h	710.7336
Reaction residence time	h	870.5983
Advection residence time	h	3870.547

Figure 7.12.   An illustrative output of a Level III computer program.

7.12   UNSTEADY STATE CONDITIONS (LEVEL IV)

It is relatively simple to extend the Level III model to unsteady state conditions. Instead of writing the steady state mass balance equations for each medium we write a differential equation. In general it takes the form for compartment i

$$V_i Z_i df_i/dt = I_i + \Sigma(D_{ji}f_j) - D_{Ti}f_i$$

where V_i is volume, Z_i is bulk Z value, I_i is the input rate (which may be a function of time), each term $D_{ji}f_j$ represents intermedia input transfer, and $D_{Ti}f_i$ is the total output. If an initial fugacity is defined for each medium, these four equations can be integrated to give the fugacities as a function of time, thus quantifying the response characteristics of the system. An example of this approach has been given by Cohen and Ryan (1985), who examined the time course of benzo-a-pyrene concentration changes in a region of the United States.

Integration is probably best done numerically using standard procedures. In principle it is also possible to obtain the analytical solution to the set of simultaneous linear differential equations but the algebra is messy when there are more than two compartments.

The new information obtained from the Level IV calculation are the response times. These times can be inferred from the characteristic times (VZ/D) of the various processes as applied to each medium; thus, it is not always necessary to undertake the integration except to confirm that the inferred values are reasonable. These response times are of considerable interest in the regulatory context because they control how long it may take to clean up a contaminated region such as a bottom sediment.

The interested reader can consult the compilation edited by Cohen (1986) for examples of multimedia models and their application to chemical exposure estimation.

8 APPLICATIONS OF FUGACITY MODELS

8.1 INTRODUCTION

The ability to define Z and various D values, then deduce fugacities, concentrations, fluxes, and amounts leads to a capability of addressing a series of environmental modeling problems which are more localized or site-specific than the general level III problem addressed in Chapter 7.

Applications of this type have been described by Thibodeaux et al. (1986) as "vignette" models on the basis of Webster's dictionary definition of vignette as "a short literary sketch, chiefly descriptive, and characterized usually by delicacy, wit and subtlety." Thibodeaux has defined the characteristics of a vignette model as simple mathematics, narrow scope, few input variables, verified physico-chemical phenomena, and ease of use and interpretation. We develop a series of these models in this chapter treating chemical behavior (a) at the air-water interface, (b) in surface soils, (c) at the sediment-water interface, (d) in ponds or lakes, (e) in rivers, (f) when bioaccumulating from water to fish, and (g) in the indoor environment. Finally, we briefly discuss the calculation of exposure of organisms to chemicals, including aspects of quantitative structure activity relationships (QSARs), and models in the pharmacological setting within an animal body.

In several cases a simple model computer program is developed and copies are provided on the diskette.

The aim is to provide the reader with a specimen calculation of chemical fate in these situations in the expectation that the parameter values describing the environment and the chemical can be modified to simulate specific situations. It may be desirable to add or delete processes or change the model structure to suit individual requirements. Many of the models apply to steady-state conditions and can be reformulated to describe time-varying conditions as differential, rather than algebraic equations. These differential equations can be solved algebraically or integrated numerically depending on their complexity.

Some of the most satisfying moments in environmental science come when a model is successfully fitted to experimental or observed data and it becomes apparent that the important chemical transport and transformation processes are being represented with fidelity. Even more satisfying is the subsequent use of the model to predict chemical fate in as yet uninvestigated situations, leading to gratifying and successful "validation." Failure of the model, while disappointing, is often instructive and exciting as a demonstration that our fundamental understanding of environmental processes is flawed, and further investigation is needed.

8.2   AN AIR-WATER EXCHANGE MODEL

8.2.1   Introduction

Air-water exchange process calculations are useful when estimating rates of chemical loss from treatment lagoons, ponds, and lakes, for estimating deposition rates of atmospheric contaminants, and for interpreting observed air and water concentrations to establish the direction and rate of transfer. The complexity of the several processes and the widely varying physical chemical properties of candidate chemicals leads to situations in which chemical behavior is not necessarily intuitively obvious.

There have been several treatments of air-water exchange in the literature, notably in the compilation by Lyman et al. (1982) and by Brutsaert and Jirka (1984), and the two volume compilation by Pruppacher et al. (1983). Most texts on air pollution contain extensive treatments of aerosol properties and deposition processes.

The simple model derived here can provide a rational method of estimating exchange characteristics and exploring the sensitivity of the results to assumed values of the various chemical and environmental parameters. A program is provided on the diskette. It can be translated into other languages, and can be modified as desired to incorporate other conditions and features of chemical behavior.

8.2.2   The Nature of the Media

The situation treated here, and the resulting model, are largely based on the study of air-water exchange by Mackay et al. (1986) and are depicted in Figure 8.1.

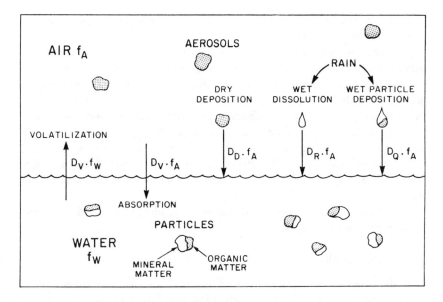

Figure 8.1.   Air-water exchange processes.

The water phase area and depth (and hence volume) are defined, it being assumed that the water is well-mixed. The water contains suspended particulate matter which may contain mineral and organic material. The concentration (mg/L or g/m³) of suspended matter is defined, as is its organic carbon (OC) content (g OC per g dry particulates). By assuming a 56% OC content of organic matter, the masses of mineral and organic matter can be deduced. Densities of 1000, 1000, and 2500 kg/m³ are assumed for water, organic matter, and mineral matter, thus enabling the volumes and volume fractions to be deduced.

If the concentration of suspended particulate matter in water is C_P g/m³, and the organic carbon content is y g OC/g dry particulates, then the organic matter content is y/0.56 g/g, and the mineral content $(1 - y/0.56)$ g/g. Introducing the densities (kg/m³) enables the volumes to be calculated as

organic matter $(y/0.56)/(1000/\rho_{OM})$ cm³/g particulate
mineral matter $(1 - y/0.56)/(1000/\rho_{MM})$ cm³/g particulate

thus the volumes of organic and mineral matter per cubic meter of water (i.e., the volume fractions) are

organic matter $C_P(y/0.56)/(1000\ \rho_{OM}) = v_{OM}$
mineral matter $C_P(1 - y/0.56)/(1000\ \rho_{MM}) = v_{MM}$

The volume fraction of water is strictly $(1 - v_{OM} - v_{MM})$, but this is usually negligibly different from unity.

The air phase is treated similarly, having the same area and a defined (possibly arbitrary) height and containing a specified concentration (ng/m³) of aerosols or atmospheric particulates. By assuming an aerosol density of 1500 kg/m³, the volume fraction of aerosols can be deduced. No information on aerosol composition, size distribution, or surface area is sought or used.

If the concentration of aerosols or total suspended particulates is TSP ng/m³, this corresponds to 10^{-12}.TSP kg/m³, and to a volume fraction v_Q of 10^{-12} TSP/ρ_Q where ρ_Q is the aerosol density (2000 kg/m³). Thus, a typical TSP of 30000 ng/m³ or 30 μg/m³ is equivalent to a volume fraction of 15 × 10^{-12}.

The phase volumes can be calculated as the product of total volume V and respective volume fractions.

8.2.3 Partitioning

In the program, the total concentrations of chemical in the air and water are requested along with physical chemical property data.

A fugacity calculation is then completed of the partitioning status of the chemical, assuming particles to be in equilibrium with the medium. For each phase a Z value is calculated from the specified physical chemical properties of the substance of interest.

For air, Z_A is calculated as $1/RT$, and for water Z_W is calculated as $1/H$, using vapor pressure, solubility, or air-water partition coefficient data as discussed in Chapter 5.

An organic carbon partition coefficient K_{OC} is used to describe water-water particle organic matter partitioning. K_{OC} may be estimated from the octanol-water partition coefficient K_{OW} using the Karickhoff (1981) correlation

$$K_{OC} = 0.41 \ K_{OW}$$

Assuming an organic carbon content of 56% in organic matter, the organic matter partition coefficient K_{OM} can be calculated as 0.56 K_{OC}. K_{OM} has dimensions of L/kg; thus, a density correction must be applied when calculating Z for organic matter, a factor of 1000 being included to correct density from kg/m³ to kg/L.

$$Z_{OM} = K_{OM} \ \rho_{OM} Z_W / 1000 = 0.56 Z_{OC} \quad \text{where } K_{OM} = 0.56 \ K_{OC}$$

A dimensionless mineral matter-water partition coefficient K_{MW} can also be defined enabling its Z value to be deduced, if desired.

$$Z_{MM} = K_{MW} \ \rho_{MM} Z_W / 1000$$

Usually mineral-water partitioning is insignificant in comparison to organic carbon-water partitioning; thus, an arbitrary low value of K_{MW} of unity may be used. If the particulate matter is of low organic content K_{MW} can become important, and measurement may be necessary.

In the water phase a total concentration C_{TW} (mol/m³) is deduced from the input concentration which is WC_{TW} (g/m³ or mg/L), W being the molecular mass of the chemical (g/mol). The total Z value for the bulk water phase is deduced by adding the component Z values in proportion to their volume fractions.

$$Z_{TW} = v_W Z_W + v_{OM} Z_{OM} + v_{MM} Z_{MM}$$

The common fugacity f_W is then C_{TW}/Z_{TW} and the various phase concentrations are calculated as Zf_W mol/m³ of phase or vZf_W mol/m³ of water.

A simpler method, which is used in the program, is to neglect the mineral matter and calculate Z_S for the bulk particles as

$$Z_S = Z_W K_{OC} y \rho_S / 1000$$

where ρ_S is the particle density (say 1500 kg/m³). The volume fraction of particles v_S is $C_P/(1000\rho_S)$ and Z_{TW} is $(v_W Z_W + v_S Z_S)$.

For the aerosol phase two approaches may be used. As was discussed in Chapter 6, experimental measurements are often made of the ratio of gaseous to particle-associated chemical by filtering an air sample prior to sorption on a polyurethane foam plug. The quantities of chemical trapped on the foam and the filter thus give the gaseous or air phase quantity to particulate or filterable quantity, designated the A/F ratio. Such ratios have been reported, as discussed earlier, and correlated as a function of temperature and chemical properties, notably subcooled liquid vapor pressure. The ratio A/F is obviously influenced by total suspended particulate (TSP ng/m³) content; thus, it is common to report and correlate the group TSP.A/F. For example, Yamasaki et al. (1982) report values of this ratio for pyrene of about 3×10^6, which at a TSP of 10^5 ng/m³ implies a 30:1, gas: particulate ratio.

The second approach is to deduce a dimensionless aerosol-air partition coefficient K_{QA}, which in this case has the advantage that Z_Q for aerosols can be deduced directly as $K_{QA}Z_A$. Here K_{QA} is a ratio of volumetric concentrations e.g., (mol/m³ aerosol)/(mol/m³ air). Mackay et al. (1986) have tentatively correlated K_{QA} as a function of subcooled liquid vapor pressure P_L^S(Pa).

$$K_{QA} = 6 \times 10^6 / P_L^S$$

For liquid chemicals the liquid vapor pressure can be used directly, but for solids the subcooled liquid vapor pressure must be estimated from the solid vapor pressure P_S^S using the fugacity ratio F, which can be estimated from the chemical's melting point T_M (K) and the data temperature T (K) by assuming an entropy of fusion of 56 J/mol K, giving

$$P_L^S = P_S^S/F \text{ where } F = \exp (6.79(1 - T_M/T))$$

The term 6.79 is $\Delta S/R$ i.e., 56/8.314, the group $T\Delta S$ being the enthalpy of vaporization.

If the air and aerosol phases are in equilibrium, a common fugacity f_A applies and is related to the concentrations as follows

$$f_A = C_A/Z_A = C_Q/Z_Q = C_{TA}/Z_{TA}$$

where C_A is the gaseous concentration (mol/m³), C_Q is the concentration on the aerosols (mol/m³ of the aerosols), C_{TA} is the total air concentration, and Z_{TA} is the bulk air Z value given by:

$$Z_{TA} = Z_A(1 - v_Q) + v_Q Z_Q \approx Z_A + v_Q Z_Q$$

The concentration on the aerosol is also $v_Q C_Q$ mol/m³ of air.

Since v_Q is $10^{-12} TSP/\rho_Q$ and the ratio F/A is $v_Q C_Q/C_A$ or $v_Q K_{QA}$ it follows that

$$TSP.A/F = 10^{12}\rho_Q/K_{QA} = 10^6 P_L^S \rho_Q/6$$

The group (TSP.A/F) is thus an alternative, indirect method of expressing the magnitude of K_{QA}. Care must be taken to check the units of TSP which are here ng/m³, but are often μg/m³. For example, pyrene with a solid vapor pressure of 6×10^{-4} Pa and a melting point of 156°C has a subcooled liquid vapor pressure of 1.2×10^{-2} Pa; thus, K_{QA} is 500×10^6, giving a group TSP.A/F of 4×10^6 which is in agreement with Yamasaki's value of 3×10^6 discussed earlier.

The program uses P_S^S and melting points to calculate P_L^S (if necessary), then calculates TSP.A/F, which can be checked for compatibility with data from other sources. No temperature correction is applied, so the P_S^S or P_L^S data must be at the required environmental temperature.

A check is desirable of the magnitude of f/P_S^S or f/P_L^S. When this ratio equals one, saturation is achieved, and when it exceeds one the chemical will "precipitate" as a pure phase, i.e., its solubility in air or water is exceeded. Normally the ratio is much less than unity.

8.2.4 Transport

Four processes are considered, as shown in Figure 8.1 (a) diffusive exchange by volatilization and the reverse absorption, (b) dry deposition of aerosols,

(c) wet dissolution of chemical, and (d) wet deposition of aerosols. In each case a D value (mol/Pa.h) is used to characterize the rate, which is Df mol/h.

For diffusion, the two-resistance approach is used, two D values being deduced, one for the air, and one for the water boundary layer, D_A and D_W as

$$D_A = k_A A Z_A \qquad D_W = k_W A Z_W$$

where k_A and k_W are mass transfer coefficients with units of m/h and A is area (m²). Illustrative values of 5 m/h for k_A and 5 cm/h for k_W can be used but it should be appreciated that environmental values can vary widely and a separate calculation may be needed as described by Mackay and Yuen (1983).

The overall resistance $(1/D_V)$ is obtained by adding the series resistances $(1/D)$ as

$$1/D_V = 1/D_A + 1/D_W$$

The rate of vaporization is then $f_W D_V$, the rate of absorption is $f_A D_V$, and the net rate of vaporization is $D_V(f_W - f_A)$. An overall mass transfer coefficient is also calculated.

For dry deposition, a dry deposition velocity U_P is used, a typical value being 0.3 cm/s or 10 m/h. The total dry deposition rate is thus $U_P v_Q A$ m³/h, the corresponding D value D_D is

$$D_D = U_P v_Q A Z_Q$$

and the rate is $D_D f_A$ mol/h

For wet dissolution a rain rate is defined, usually in units of m/year, a typical value being 0.5 m/year or 6×10^{-5} m/h, designated U_R. The total rain rate is then $U_R A$ m³/h and the D value D_R is

$$D_R = U_R A Z_W$$

and the rate is $D_R f_A$ mol/h.

For wet aerosol deposition a scavenging ratio Q is used representing the volume of air efficiently scavenged by rain of its aerosol content, per unit volume of rain. A typical value for Q of 200,000 may be used. The volume of air scavenged per hour is thus $U_R A Q$ m³/h which will contain $U_R A Q v_Q$ m³/h of aerosol, thus the D value D_Q is

$$D_Q = U_R A Q v_Q Z_Q$$

and the rate is $D_Q f_A$ mol/h.

A washout ratio is often employed in such calculations. This is the dimensionless ratio of rain to total air concentration either on a volumetric (g/m³ rain per g/m³ air) or on a gravimetric (mg/kg per mg/kg) basis. The total rate of chemical deposition in rain is $(D_R + D_Q)f_A$; thus, the concentration in the rain is $(D_R + D_Q)f_A/U_R A$ or $f_A(Z_W + Qv_Q Z_Q)$ mol/m^3. The total air concentration is $f_A Z_{TA}$; thus, the volumetric washout ratio is $(Z_W + Qv_Q Z_Q)/(Z_A + v_Q Z_Q)$. The gravimetric ratio is smaller by the ratio of air to water densities i.e., approximately 1.2/1000. If the chemical is almost entirely aerosol-associated, as is the case with metals such as lead, the volumetric washout ratio approaches Q. These washout ratios are calculated and can be used to check against reported values.

The total rates of transfer are thus

water to air $f_W.D_V$ mol/h
air to water $f_A(D_V + D_D + D_R + D_Q) = f_A D_T$ mol/h

The total amounts of chemical in each phase may be calculated as $V_A Z_{TA} f_A$ and $V_W Z_{TW} f_W$, thus, the rate constants (h^{-1}) and half-times (h) for transfer from each phase are

From air $D_T/(V_A Z_{TA})$ h^{-1} and $0.693 V_A Z_{TA}/D_T$ h
From water $D_V/(V_W Z_{TW})$ h^{-1} and $0.693 V_W Z_{TW}/D_V$ h

These quantities are useful as indicators of the rapidity with which chemical can be cleared from one phase to the other, thus enabling the significance of these exchange processes to be assessed relative to other processes such as reaction. Inspection of the D values shows which processes are most important.

It is noteworthy that a steady state (i.e., no net transfer) condition may apply in which the air and water fugacities are unequal, i.e., a nonequilibrium, steady state applies. The steady state will apply when

$$f_W D_V = f_A D_T$$

The steady state water fugacity and concentration with respect to the air, and the steady state air fugacity and concentration with respect to the water can thus be calculated to give an impression of the extent to which the actual concentrations depart from the steady state values, as distinct from the equilibrium (equi-fugacity) values.

8.2.5   Model

The supplied computer program in BASIC language contains sufficient comments to enable the logic to be followed. The program, as written, allows for specification of a chemical by the user, one already in the database, or (when doing repeated calculations), the previously specified chemical. Similarly, the air and water conditions and chemical concentrations can be selected, or illustrative values used. For convenience, previously specified values are printed to enable sensitivity analyses to be conducted. The program interrogates the user for concentrations in air and water, zero being an acceptable input. It then calculates the fluxes and the steady state fugacities in each phase corresponding to the specified concentration in the other phase. The nature of the output is self-explanatory.

8.3 A SURFACE SOIL MODEL

8.3.1 Introduction

Chemicals are frequently encountered in surface soils as a result of deliberate application of agrochemicals and inadvertent spillage and leakage. It is often useful to assess the likely fate of the chemical, i.e., how fast the rates of degradation, volatilization, and leaching in water are likely to be, and how long it will take for the soil to "recover" to a specified, or acceptable, level of contamination. Remedial measures such as excavation may be needed in cases in which recovery time is unacceptably long. Persistence is also an important characteristic of pesticide selection and application.

Most modeling efforts in this context have been for agrochemical purposes, the most comprehensive recent effort being described in a series of publications by Jury, Spencer and Farmer(1983, 1984a, 1984b, 1984c). Other notable models are reviewed in these papers. The present model is essentially a very simplified version of the Jury model (1983) and is a modification of a published herbicide fate model (Mackay and Stiver 1991).

The general issue of chemical fate in soils has been treated in several texts; for example, Sawhney and Brown (1989) and Sposito (1989). A convenient, brief account of soil classification systems is given by Green (1988).

8.3.2 Soil Composition

The soil matrix illustrated in Figure 8.2 is treated as four phases: air, water, organic matter, and mineral matter, subscripted 1,2,3,and 4, respectively. The organic matter is considered to be 56% organic carbon. The volume fractions of air and water are defined, either by the user or by default values, as is the organic carbon (OC) content on a g OC per g dry soil basis. Assuming densities of 1.19, 1000, 1000 and 2500 kg/m^3 for air, water, organic matter, and mineral matter permits the mass and volume fractions of each phase, and the overall soil density to be calculated.

The soil area and depth are specified, thus enabling the total volumes and mass of soil and its component phases to be deduced. The amount of chemical

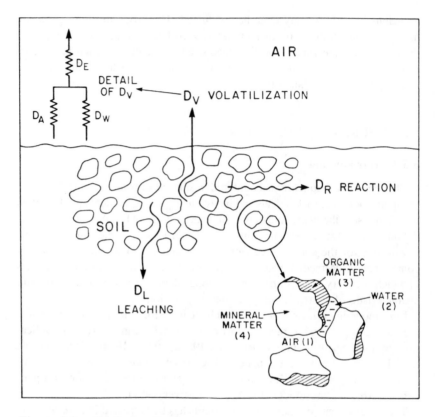

Figure 8.2.   Surface soil processes.

present is specified in units of kg/ha, which is convenient for agrochemicals, from which the total amount of chemical present is deduced. The chemical is assumed to be homogenously distributed throughout the entire soil volume.

8.3.3   Partitioning

A fugacity calculation is then completed of the partitioning status of the chemical in the soil using physical chemical property data. For each phase a Z value is calculated starting with Z_1 for air and Z_2 for water as described earlier. Occasionally vapor pressure data for pesticides are available as vapor densities, thus a check can be made to ensure that values of vapor pressure and density are consistent, and that units have been correctly interpreted.

For organic matter, an organic carbon-water partition coefficient K_{OC} may be estimated from the octanol water partition coefficient K_{OW} using the

Karickhoff (1981) correlation that K_{OC} is 41% of K_{OW}. Alternatively, a reported K_{OC} may be used. From K_{OC}, K_{OM} is calculated as 0.56 K_{OC}, i.e., the organic matter is 56% organic carbon.

K_{OM} has dimensions of L/kg; thus, a density correction must be applied when calculating Z_3 as

$$Z_3 = K_{32}\rho_3 Z_2/1000$$

The factor of 1000 converts density from kg/m³ to kg/L.

A mineral matter-water partition coefficient K_{MW} or K_{42} with units of L/kg is defined by the user, enabling Z_4 for mineral matter to be deduced.

$$Z_4 = K_{42}\rho_4 Z_2/1000$$

The magnitude of K_{42} is often in doubt. It varies with water content of the soil, being generally larger for very dry soils. In the absence of other information, a value of about one can be assumed for illustrative purposes. It is suspected that much of the mineral matter is covered with detrital organic matter and is not readily accessible to the potentially sorbing chemical. For hydrophobic chemicals the contribution of mineral matter to the total sorptive capacity is usually small and can be neglected. Exceptions occur when the OC content is very low, i.e., less than 0.1%, or for very dry soils. There is no substitute for direct measurement of K_{42}, and indeed K_{32}, for the soil of interest.

At equilibrium, a common fugacity f applies to all soil phases; thus, the total amount of chemical present, M mols, will be distributed at concentrations C_1 to C_4 (mol/m³), as described by a Level I calculation.

$$M = C_1V_1 + C_2V_2 + C_3V_3 + C_4V_4$$
$$\text{but } C_i = Z_i f$$
$$\text{thus } M = f(Z_1V_1 + Z_2V_2 + Z_3V_3 + Z_4V_4)$$

from which f can be deduced, followed by the concentrations C_i, the amounts m_i(mols) as C_iV_i, the percentage amounts 100 m_i/M, and concentrations in other units such as g/m³ and μg/g. An overall or bulk soil Z value, Z_{ST}, can also be calculated either as the sum of the volume fraction, phase Z value products, or as total concentration (mol/m³)/fugacity. The fugacity is then C_{ST}/Z_{ST} where C_{ST} is the total concentration, (mol/m³ soil).

It is again prudent to examine the fugacity to check that it is less than the vapor pressure. If it exceeds the vapor pressure, phase separation of pure chemical will occur, i.e., the capacity of all phases to "dissolve" chemical is exceeded. This can occur in heavily contaminated soils which have been subject to spills, or when there is heavy application of a pesticide. Essentially, the "solubility" of the chemical in the soil is exceeded.

This partitioning behavior provides an insight into the amounts present in the air and water phases and thus subject to migration or diffusion. It also shows the extent to which organic matter dominates the sorptive capacity of the soil.

8.3.4 Transport and Transformation

Three processes are considered: degrading reactions, volatilization, and leaching. In each case the rate is characterized by a D value.

An overall reaction half-life is specified t (h) from which an overall rate constant $k_R(h^{-1})$ is deduced as $0.693/t$. The reaction D value D_R is then calculated from the total soil volume V_T and Z value Z_T as

$$D_R = k_R V_T Z_T$$

In principle, if a rate constant k_i is known for a specific phase, the phase-specific D value can be deduced as $k_i V_i Z_i$, but normal practice is to report an overall rate constant applicable to the total amount of chemical in the entire soil matrix. If no reaction occurs it is best to insert a fictitiously large value for the half-life.

A water leaching rate is specified in units of mm/day. This may represent rainfall (which is typically 1 to 2 mm/day) or irrigation. This rate is converted into a total water flow rate G_L m³/h, which is combined with the water Z value to give the leaching D value.

$$D_L = G_L Z_2$$

This assumes that the concentration of chemical in the water leaving the soil is equal to that in the water in the soil, i.e., local equilibrium has become established and no bypassing or "short circuiting" occurs. The "solubilizing" effect of dissolved or colloidal organic matter in the soil water is ignored, but it could be included by increasing the Z value of the water to account for this extra capacity.

Volatilization is treated using the approach suggested by Jury et al. (1983). Three contributing D values are deduced.

An air boundary layer D value D_E is deduced as the product of area A, a mass transfer coefficient k_V and the Z value of air.

$$D_E = A k_V Z_1$$

Jury has suggested that k_V be calculated as the ratio of the chemical's molecular diffusivity in air (0.43 m²/day or 0.018 m²/h being a typical value), and an air boundary layer thickness of 4.75 mm (.00475 m); thus, k_V is typically 3.77 m/h. Another k_V value may be selected to reflect different micrometerological conditions.

An air-in-soil diffusion D value is deduced to characterize the rate of transfer of chemical through the soil in the air phase. Following Jury, the Millington-Quirk equation is used to deduce an effective diffusivity B_{EA} from the air phase molecular diffusivity B_A as follows:

$$B_{EA} = B_A . v_1{}^{10/3}/(v_1 + v_2)^2$$

where v_1 is the volume fraction of air and v_2 is the volume fraction of water. If v_2 is small, this reduces to a dependence on v_1 to the power 1.33. A diffusion path length Y must be specified, which is the vertical distance from the position of the chemical of interest to the soil surface, i.e., it is not the "tortuous" distance.

The air diffusion D value D_A is then:

$$D_A = B_{EA} . A . Z_1/Y$$

A similar approach is used to calculate the D value for chemical diffusion in the water phase in the soil, except that the molecular diffusivity in water B_W is used, a value of 4.3×10^{-5} m²/day being assumed and the water volume fraction and Z value being used, namely:

$$D_W = B_{EW}AZ_2/Y$$
$$\text{where } B_{EW} = B_W v_2{}^{10/3}/(v_1 + v_2)^2$$

Since the diffusion D values D_A and D_W apply in parallel, the total D value for chemical transfer from bulk soil to soil surface is $(D_A + D_W)$. The boundary layer D value then applies in series so that the overall volatilization D value D_V is given by:

$$1/D_V = 1/D_E + 1/(D_A + D_W)$$

This is illustrated in Figure 8.2.

Selection of the diffusion path length Y involves an element of judgment. If, for example, chemical is equally distributed in the top 20 cm of soil, an average value of 10 cm for Y may be appropriate as a first estimate. This will greatly underestimate the volatilization rate of chemical at the surface. Since the rate is inversely proportional to Y, a more appropriate selection of a single value of Y as the average applying between two depths Y_1 and Y_2 is the log mean of Y_1 and Y_2, i.e., $(Y_1 - Y_2)/\ln(Y_1/Y_2)$. Unfortunately a zero (surface) value of Y can not be used when calculating the log mean. For chemical between depths of 1 and 10 cm, a log mean depth of 3.9 cm is more appropriate than the arithmetic mean of 5.5 cm. It may be useful to consider layers of soil separately, e.g., 2 to 4 cm, 4 to 6 cm, etc., and calculate separate volatilization rates for each. Deeper chemical will thus volatilize more slowly, leaving the remaining chemical more susceptible to removal by other processes.

It is acceptable to specify a mean Y of, say, 10 cm to examine the fate of chemical in the 2 cm depth region from 9 to 11 cm, i.e., Y can exceed the contaminated depth. This depth issue is irrelevant to reaction or leaching, but it must be appreciated that if the soil is treated as separate layers, the leaching rate is applicable to the total soil, not to each layer independently.

The total rate of chemical removal is then $f.D_T$ where the total D_T value is given by

$$D_T = D_R + D_L + D_V$$

the individual rates being $f.D_R$, $f.D_L$ and $f.D_V$. The overall rate constant k_O is thus D_T/V_TZ_T where V_TZ_T is the sum of the V_iZ_i products, and the overall half-life t_O is $0.693/k_O$ hours. The half-life attributable to each process individually t_i is $0.693D_i/V_TZ_T$ thus

$$1/t_O = 1/t_R + 1/t_L + 1/t_V$$

It is illuminating to calculate the rates of each process, the percentages, and the individual half-lives. Obviously the shorter half-lives dominate. The situation being simulated is essentially the first order decay of chemical in the soil by three simultaneous processes; thus, the amount remaining from an initial amount M mols at any time t hours will be:

$$M \exp(-D_Tt/VZ) = M \exp(-k_Ot) \text{ mols}$$

This relatively simple calculation can be used to assess the potential for volatilization, or for groundwater contamination.

Implicit in this calculation is the assumption that the chemical concentration in the air and in the entering leaching water are zero. If this is not the case, an appropriate correction must be included. In principle, it is possible to estimate atmospheric deposition rates as was done in Section 8.2, and couple these processes to the soil fate processes in a more comprehensive air-soil exchange model.

8.3.5 Model

The supplied computer program in BASIC language is similar in structure to the air-water model.

Users are encouraged to modify the various parameter values, and are cautioned that the values given are not necessarily widely applicable. It should be noted again that varying the input temperature will not vary physical chemical properties such as vapor pressure. Temperature dependence must be entered "by hand." Other chemicals can be substituted or added as desired by modifying the appropriate lines of coding.

8.4   A SEDIMENT-WATER EXCHANGE MODEL

8.4.1   Introduction

Exchange of chemical at the sediment-water interface can be important for the estimation of (a) the rates of accumulation or release from sediments; (b) the concentration of chemicals in organisms living in, or feeding from the benthic region; (c) which transfer processes are most important in a given situation; and (d) in the case of "in-place" sediment contamination, the likely recovery times. The complexity of the system and the varying properties of chemicals of possible concern leads to a situation in which a specific chemical's behavior is not necessarily obvious.

8.4.2   The Nature of the Media and the Processes

The situation treated here, and the resulting model, are largely based on a discussion of sediment-water exchange by Reuber et al. (1987), by Eisenreich (1987), and in part on a report by Formica et al. (1988). It is depicted in Figure 8.3.

The water phase area and depth (and hence volume) are defined, it being assumed that the water is well mixed. The water contains suspended particulate matter which may contain mineral and organic material. The concentration (mg/L or g/m³) of suspended matter is defined, as is its organic carbon (OC) content (g OC per g dry particulates). The volume fractions are calculated similarly to those in Section 8.2 for air-water exchange.

The sediment phase is treated similarly, having the same area, a defined well-mixed depth and a specified concentration of solids and interstitial or pore water. The organic carbon content of the sediment solids is defined, from which an organic matter content is deduced, again assuming organic matter to be 56% organic carbon. The colloidal or particulate organic carbon content of the interstitial water can be included (but is not) if it is suspected that these colloids play an important role in enhancing chemical diffusion.

Rates of sediment deposition, resuspension, and burial are specified, as are first order reaction rates in the sediment phase. Allowance could be (but is not) made for infiltration of ground water through the sediment in either (but not both) vertical directions. Lipid contents of organisms present in the water and sediment are specified for later illustrative bioconcentration calculations.

Partitioning

A fugacity partitioning calculation is completed for the water and sediment phases using specified total chemical concentrations, (g/m³ or mg/L) in the

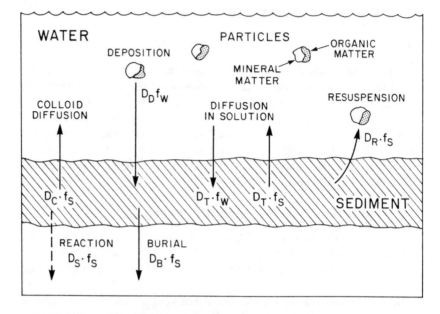

Figure 8.3.   Sediment-water exchange processes.

water, and µg/g of dry sediment solids in the sediment. This requires input
of chemical properties of molecular mass, water solubility, vapor pressure,
and octanol-water partition coefficient. For each phase a Z value is calculated,
as in earlier sections. Since no air phase appears in the calculation, the vapor
pressure is not strictly necessary. Identical concentration, but not fugacity,
results are obtained when an arbitrary vapor pressure is used.

An organic carbon partition coefficient K_{OC} is used to describe partitioning between water and organic matter in suspended particles and sediment
solids. K_{OC} and K_{OM} are again estimated from the octanol-water partition
coefficient K_{OW} using the Karickhoff correlation.

In the water phase a total concentration C_{TW} (mol/m^3) or WC_{TW} (g/m^3) is
defined, W being the molecular mass of the chemical (g/mol). The total Z
value for the bulk water phase is deduced by adding the component Z values
in proportion of their volume fractions.

$$Z_{TW} = v_W Z_W + v_{OM} Z_{OM}$$

The common fugacity f_W is then C_{TW}/Z_{TW} and the various phase concentrations are, as before,

water $C_W = Z_W f_W$

organic matter $C_{OM} = Z_{OM} f_W$ mol/m^3 organic matter

or $v_{OM} Z_{OM} f_W$ mol/m^3 water

Partitioning to mineral matter could be included, and is so in the program.

The sediment phase is assumed to consist of solids which contain a defined organic carbon content, and pure pore water. The commonly measured concentration is that of chemical on the solid phase, (i.e., chemical in pore water or colloids in the pore water is not included). This concentration in units of μg/g is defined from which the volumetric concentration on the solids (C_S mol/m^3) is deduced. Z values are deduced for the solids (Z_S), the interstitial water (Z_W), and the bulk sediment Z_{TS} phase by adding the component Z values in proportion of the volume fractions, v_W for water and v_S for solids.

$$Z_{TS} = v_W Z_W + v_S Z_S$$

Since C_S and Z_S are known, the fugacity f_S can be deduced as C_S/Z_S. Assuming the pore water to be in equilibrium with the solids, the concentration in this water can be deduced, as can the bulk concentration.

Illustrative biotic Z values can be deduced for both water and sediment as $K_B \cdot Z_W$, where the bioconcentration factor K_B is estimated from the product of lipid content L_B (e.g., 0.05) and K_{OW}, i.e., $L_B K_{OW}$. Biota are included for illustrative purposes and are not used in the mass balance.

It is possible to modify the model to include colloidal matter present in the pore water. This could be done by defining a Z_C for colloidal matter and a volume fraction v_C and adding $v_C Z_C$ to the equation for Z_{TS}. This would increase the concentration in the bulk pore water. It is likely that diffusion of this colloidal matter, and its associated chemical, is an important process.

The total and contributing concentrations in all phases and the fugacities can thus be deduced. From the biotic Z values the corresponding concentrations can also be deduced in water and sediment.

Transport

Five processes are considered: (a) sediment deposition, (b) sediment resuspension, (c) sediment burial, (d) diffusive exchange of water between the water column and the pore water, and (e) sediment reaction. Each is expressed as a D value.

The deposition D value D_D is calculated as the product of volumetric deposition rate G_D m^3/h and the particle Z value Z_P

$$D_D = G_D Z_P$$

The resuspension D value D_R is calculated similarly as $G_R Z_S$

$$D_R = G_R Z_S$$

The burial D value D_B is calculated as

$$G_B Z_S$$

For diffusive exchange of water, the D value D_T is calculated from an overall water phase mass transfer coefficient (MTC) k_W, the area A and the Z value for water

$$D_T = k_W A Z_W$$

The mass transfer coefficient can be calculated from an effective diffusivity, corrected for the sediment solids content as discussed by Formica et al. (1988), and a path length.

For reaction the overall rate constant is k_R and the D value is

$$D_S = V_S Z_{TS} k_R$$

The individual and total rates of transfer can be calculated as the D.f products. The overall steady-state mass balance is thus

$$f_W (D_D + D_T) = f_S (D_R + D_T + D_B + D_S)$$

from which the steady-state water and sediment fugacities corresponding to the defined sediment and water fugacities can be deduced. Response times can also be calculated for each medium.

It is noteworthy that, as in the air-water example, a steady state condition is reached of unequal water and sediment fugacities. This is because the non-diffusive deposition process is generally faster than the resuspension process. Diffusion always struggles to restore equilibrium, but it can proceed only slowly, especially for hydrophobic chemicals.

8.4.3   Model

The program, as given on the diskette, allows for specification of sediment and water conditions and chemical concentrations, or illustrative values can be used. The program calculates the fluxes and the steady-state fugacity in each phase corresponding to the specified concentration in the other phase. The program output is self-explanatory.

Users are encouraged to modify the various parameter values and are cautioned that the values given are not necessary widely applicable. Again, varying temperature will not vary physical chemical properties.

8.5   QWASI MODEL OF CHEMICAL FATE IN LAKES

8.5.1   Introduction

Having established air-water and sediment-water exchange models, it is relatively straightforward to combine them in a lake model by adding reaction and advective inflow and outflow terms. This was done when developing the QWASI (Quantitative Water Air Sediment Interaction) model by Mackay et al. (1983a, c) and has since been applied to Lake Ontario (Mackay 1989) and to ponds (Southwood et al. 1989). In principle, it can be applied to any well mixed-body of water for which the hydraulic and particulate flows are defined.

Figure 8.4 shows the transport and transformation processes treated, and Table 8.1 lists the D values and the corresponding rate expressions. It is convenient to group the D values as shown. No colloid transfer is included, but if desired it can be included in the resuspension D values. Table 8.2 gives the mass balance equations in unsteady state or differential form, and in steady state form. The steady state solution describes conditions which will be reached after prolonged exposure of the lake to constant input conditions, i.e., emissions, air fugacity, and inflow fugacity. Also given in Table 8.2 is the solution to the differential equations from a defined initial condition, assuming that the input terms remain constant with time. If it is desired to vary these inputs or any other terms as a function of time, the equations must be solved numerically.

Since D values add, it is clear by mere inspection which are most important and control the overall chemical fate. For example, if D_V greatly exceeds D_Q, D_C, and D_M, it is apparent that most transfer from air is by absorption. The relative magnitudes of the processes of removal from water are particularly interesting. These occur in the denominator of the $f_{W\infty}$ equation as volatilization (D_V), reaction (D_W), water outflow (D_J), particle outflow (D_Y), and a term describing net loss to the sediment. The gross loss to the sediment is (D_D + D_T) but only a fraction of this (D_S + D_B)/(D_R + D_T + D_S + D_B) is retained in the sediment, the remaining fraction (D_R + D_T)/(D_R + D_T + D_S + D_B) being returned to the water. The three terms in the numerator of the $f_{W\infty}$ equation give the inputs from emissions, inflow and transfer from air.

When the equations are solved it is relatively simple to calculate concentrations, amounts, and fluxes. An illustration of such an output is given in Figure 8.5 for PCBs in Lake Ontario (Mackay 1989). Such mass balance diagrams clearly show which processes are most important for the chemical of interest.

8.5.2   QWASI Model

A computer program is provided which processes the various Z values, volumes, areas, flows, D values, and the input parameters to give a steady state

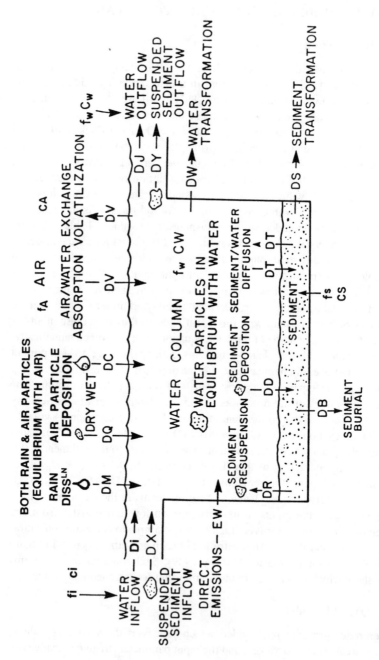

Figure 8.4. QWASI processes.

Table 8.1.   QWASI D Values and Groups as Used in the Program.

Process	D Value	Rate (mol/h)
Sediment burial	DB	GB.CS or (GB.ZS).FS or DB.FS
Sediment transformation	DS	VS.CS.KS or (VS.ZS.KS).FS or DS.FS
Sediment resuspension	DR	GR.CS or (GR.ZS).FS or DR.FS
Sediment to water diffusion	DT	KT.AS.CS/KSW or (KT.AS.ZW).FS or DT.FS
Water to sediment diffusion	DT	KT.AS.CW or (KT.AS.ZW).FW or DT.FW
Sediment deposition	DD	GD.CP or (GD.ZP).FW or DD.FW
Water transformation	DW	VW.CW.KW or (VW.ZW.KW).FW or DW.FW
Volatilization	DV	KV.AW.CW or (KV.AW.ZW).FW or DV.FW
Absorption	DV	KV.AW.CA/KAW or (KV.AW.ZW).FA or DV.FA
Water outflow	DJ	GJ.CW or (GJ.ZW).FW or DJ.FW
Water particle outflow	DY	GY.CP or (GY.ZP).FW or DY.FW
Rain dissolution	DM	GM.CA/KAW or (GM.ZW).FA or DM.FA
Wet particle deposition	DC	GC.CQ or (GC.ZQ).FA or DC.FA
Dry particle deposition	DQ	GQ.CQ or (GQ.ZQ).FA or DQ.FA
Water inflow	DI	GI.CI or (GI.ZW).FI or DI.FI
Water particle inflow	DX	GX.CX or (GX.ZP).FI or DX.FI
Direct emissions	—	EW

Nomenclature and explanation:

The groups in parentheses are the D values, i.e., DB is (GD.ZS). The rate is the product of D and fugacity F, i.e., DB.FS.

G values are flows (m³/h) of a phase, e.g., GB is m³/h, of sediment that is buried.

C values are concentrations (mol/m³), the second letter being S sediment, W water, A air, Q aerosol, P water particles, I water inflow, X water particle inflow, J water outflow, and Y particle outflow.

FW, FS, FA, and FI are fugacities of water, sediment, air and water inflow.

Z values are fugacity capacities (mol/m³.Pa), the second letter being defined as for concentration.

KS and KW are sediment and water transformation rate constants (h⁻¹).

KSW, i.e., ZS/ZW, is a sediment-water partition coefficient (dimensionless).

KAW, i.e., ZA/ZW, is an air-water partition coefficient (dimensionless).

KT is a sediment-water mass transfer coefficient and KV an overall (water-side) air-water mass transfer coefficient (m/h).

AW and AS are air-water and water-sediment areas (m²).

VW and VS are water and sediment volumes (m³).

Table 8.2. Mass Balance Equations

Sediment differential equation

$$V_S Z_{BS} df_S/dt = f_W(D_D + D_T) - f_S(D_R + D_T + D_S + D_B)$$

Water differential equation

$$V_W Z_{BW} df_W/dt = E_W + f_f(D_I + D_X) + f_A(D_V + D_Q + D_C + D_M) + f_S(D_R + D_T) - f_W(D_V + D_W + D_J + D_Y + D_D + D_T)$$

Z_{BS} and Z_{BW} are bulk phase Z values i.e. water plus solids

Steady State Solutions (i.e. derivatives equal zero)

$$f_{W\infty} = \frac{E_W + f_f(D_I + D_X) + f_A(D_V + D_Q + D_C + D_M)}{D_V + D_W + D_J + D_Y + (D_D + D_T)(D_S + D_B)/(D_R + D_T + D_S + D_B)}$$

$$f_{S\infty} = f_W(D_D + D_T)/(D_R + D_T + D_S + D_B)$$

Unsteady State Analytical Solutions

This pair of equations can be written more compactly as

$$df_W/d_t = I_1 + I_2 f_S - I_3 f_W$$
$$df_S/dt = I_4 f_W - I_5 f_S$$

where $I_1 = [E_W + f_f(D_I + D_X) + f_A(D_V + D_Q + D_C + D_M)]/V_W Z_{BW}$
$I_2 = (D_R + D_T)/V_W Z_{BW}$
$I_3 = (D_V + D_W + D_J + D_Y + D_D + D_T)/V_W Z_{BW}$
$I_4 = (D_D + D_T)/V_S Z_{BS}$
$I_5 = (D_R + D_T + D_S + D_B)/V_S Z_{BS}$

The solution with the initial conditions f_{SO} and f_{WO} is

$$f_W = f_{W\infty} + I_8 exp(-(I_6 - I_7)t) + I_9 exp(-(I_6 + I_7)t)$$
$$f_S = f_{S\infty} + \{(I_3 - I_6 + I_7)I_8 exp(-(I_6 - I_7)t) + (I_3 - I_6 - I_7)I_9 exp(-(I_6 + I_7)t)\}/I_2$$

where $f_{W\infty} = I_1 I_5/(I_3 I_5 - I_2 I_4)$ (as above)
$f_{S\infty} = I_1 I_4/(I_3 I_5 - I_2 I_4)$ (as above)
$I_6 = (I_3 + I_5)/2$
$I_7 = ((I_3 - I_5)^2 + 4I_2 I_4)^{0.5}/2$
$I_8 = \{-I_2(f_{S\infty} - f_{SO}) + (I_3 - I_6 - I_7)(f_{W\infty} - f_{WO})\}/2I_7$
$I_9 = \{I_2(f_{S\infty} - f_{SO}) - (I_3 - I_6 + I_7)(f_{W\infty} - f_{WO})\}/2I_7$

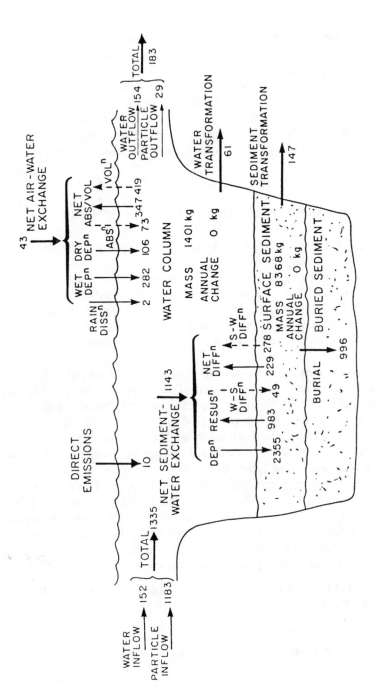

Figure 8.5.   Illustrative QWASI model steady state output for PCBs in Lake Ontario (Mackay, 1989).

algebraic solution. Algebraic unsteady state and numerical unsteady state solutions can be written by the user. This program is not as "user friendly" as the earlier vignette models. The user must specify conditions by editing the appropriate lines. The conditions simulated in the program are similar to that described by Mackay (1989) for the fate of PCBs in Lake Ontario.

Some users find it more convenient to write the equations into a spreadsheet such as Lotus 1-2-3, from which the results can be plotted directly. This form of program is often more convenient for the user, but is more difficult to transmit from user to user.

8.6 QWASI MODEL OF CHEMICAL FATE IN RIVERS

The QWASI lake equations can be modified to describe chemical fate in rivers by one of two methods.

The river can be treated as a series of connected lakes, each of which is assumed to be well-mixed with unique water and sediment concentrations. There can be varying discharges into each segment. Tributaries can be introduced as desired. The larger the number of segments, the more closely is the true "plug flow" condition of the river simulated. Figure 8.6 illustrates the approach.

The second approach is to set up and solve the differential equation for water concentration as a function of river length. This has been discussed by Mackay, Paterson, and Joy (1983b) and an application to surfactant decay in a river has been described by Holysh et al. (1985).

When setting up this equation it is preferable to use horizontal area-specific D values, designated B values (and not to be confused with the diffusivities in Chapter 7). A D value is then B.A or B.W.L where A is area, W is river width, and L is river length. The flow D values D_I and D_X are not altered in this way. Further, D_J and D_Y, the outflow D values, must equal the inflow values (unless there is loss or gain of water or particles). The emission rate e is also defined on an area basis, i.e., $mol/h.m^2$. The horizontal area of a segment is $W\Delta L$, where ΔL is the segment length.

The steady state equation for a water segment is

$$(f_W - \Delta f_W)(D_I + D_X) + e_W W\Delta L + f_A W\Delta L(B_V + B_Q + B_C + B_M) + f_S W\Delta L(B_R + B_T) = f_W W\Delta L(B_T + B_D + B_W + B_V) + f_W(D_I + D_X)$$

For a sediment segment it is

$$f_W W\Delta L(B_T + B_D) = f_S W\Delta L(B_B + B_S + B_R + B_T)$$
$$\text{thus } f_S = f_W(B_T + B_D)/(B_B + B_S + B_R + B_T)$$

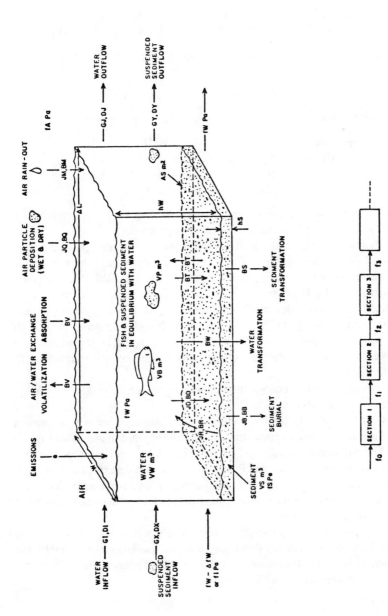

Figure 8.6.   QWASI river model.

and

$$\Delta f_W/\Delta L = [e_W W + f_A W(B_V + B_Q + B_C + B_M) -$$
$$f_W W(B_W + B_V + (B_T + B_D)(B_B + B_S)/(B_B + B_S + B_R + B_T))]/(D_I + D_X)$$

Equating $\Delta f_W/\Delta L$ to df_W/dL and integrating from an initial fugacity f_{WO} at L of zero gives

$$f_W = f_{W\infty} + (f_{WO} - f_{W\infty})\exp(-k_I L)$$

where the ultimate (L = ∞) water fugacity $f_{W\infty}$ is given by

$$f_W = [e_W + f_A(B_V + B_Q + B_C + B_M)]/[B_W + B_V +$$
$$(B_T + B_D)(B_B + B_S)/(B_B + B_S + B_R + B_T)]$$
$$\text{and } k_I = W[B_W + B_V + (B_T + B_D)(B_B + B_S)/$$
$$(B_B + B_S + B_R + B_T)]/(D_I + D_X)$$

The terms B_W, B_V, etc., in the expression for k give the relative importance of reaction, volatilization, and sediment loss processes.

It is noteworthy that this equation reduces to the oxygen uptake equation discussed earlier in Chapter 7 if all B values except B_V are zero and e_W is zero. The ultimate fugacity $f_{W\infty}$ is then f_A and k_I becomes WB_V/D_I. The group $k_I L$ is then WLB_V/D_I. Substituting $k_L Z_W$ for B_V, (where k_L is the oxygen liquid phase mass transfer coefficient) and $G_I Z_W$ or $UWYZ_W$ for D_I (where G_I is the river flowrate m^3/h, U is velocity m/h and Y is depth m) and Ut for L (where t is time), it is clear that $k_I L$ is also $k_L t/Y$ or $k_2 t$, where k_2 is the oxygen reaeration constant and is essentially the ratio of mass transfer coefficient to depth.

Oxygen uptake thus occurs with a characteristic length of $1/k_I$ or UY/k_L, or a flow time of $1/k_2$ or Y/k_L.

This expression is encountered in the oxygen sag equation which contains an additional term for oxygen consumption by organic matter added to the river. The differential equation in flow time, i.e., for a parcel of flowing water, was first written by Streeter and Phelps in 1925 and is described in texts such as that by Thibodeaux (1979). This is historically significant as probably the first successful application of a mathematical model to the fate of a chemical (oxygen) in the environment.

The key point is that by compiling D or B values it is possible to estimate chemical fate in continuously flowing systems such as rivers, thus quantifying the extent or severity of contamination at various distances or flow times downstream. No computer model is provided.

8.7   A FISH BIOCONCENTRATION MODEL

8.7.1   Introduction

The fish bioaccumulation phenomenon is very important as a means by which chemicals present at low concentration in water become concentrated by many orders of magnitude, thus causing potential hazards to the fish and to other creatures, especially birds and humans, who consume these fish. For example, DDT may be found in fish at concentrations a million times that of water. The primary cause of this effect is simply the difference in Z values between water and fish lipids as characterized by K_{ow}, but there are also other, more subtle effects at work. The kinetics of uptake are also important because a fish may never reach thermodynamic equilibrium. There is also a fascinating biomagnification phenomenon which occurs in food chains and is not yet fully understood.

It is useful to define the terminology, although opinions differ on the correct usage. Bioconcentration refers here to uptake from water. Biomagnification is the increase in concentration from food to fish. Bioaccumulation is the total (water plus food) uptake process.

In this section we address these issues, treating first uptake from water, then metabolism, growth, food uptake, and finally, food chain effects. This treatment is an abbreviated version of the study by Clark et al. (1990). Figure 8.7 depicts the processes treated.

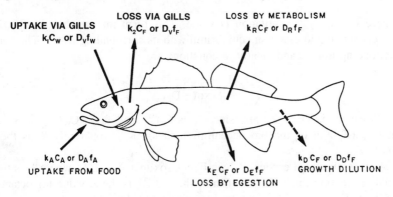

Figure 8.7.   Fish bioaccumulation processes.

8.7.2   Uptake of Chemical from Water (Bioconcentration)

The conventional concentration expression for uptake of chemical by fish from water, through the gills, was first written by Neely (1979) as

$$dC_F/dt = k_1 C_W - k_2 C_F$$

where C_F and C_W are concentrations in fish and water, k_1 is an uptake rate constant, and k_2 the clearance rate constant. Apparently the chemical passively diffuses into the fish along much the same route as oxygen. In the laboratory it is usual to expose a fish to a constant water concentration for a period of time during which the concentration in the fish should rise from zero to C_F according to the integrated version of the differential equation, with C_F initially zero, and C_W constant.

$$C_F = (k_1/k_2)C_W(1 - \exp(-k_2 t))$$

At long times C_F approaches $(k_1/k_2)C_W$ or $K_B C_W$ where K_B is the bioconcentration factor. The fish is then placed in clean water and loss or clearance or depuration is followed, the corresponding equation being

$$C_F = C_{FO} \exp(-k_2 t)$$

where C_{FO} is the concentration at the start of clearance.

We can rewrite these equations in equivalent fugacity form as

$$V_F Z_F df_F/dt = D_V(f_W - f_F)$$

where D_V is a gill ventilation D value. This form implies that the fish is merely seeking to establish equilibrium with its surrounding water. The corresponding uptake and clearance equations are

$$f_F = f_W(1 - \exp(-D_V t/V_F Z_F))$$
$$f_F = f_{FO} \exp(-D_V t/V_F Z_F)$$

Clearly k_2 is $D_V/V_F Z_F$ and since f_F is C_F/Z_F and f_W is C_W/Z_W, k_1/k_2 is merely (Z_F/Z_W) or the bioconcentration factor K_B. Apparently k_1 is then $D_V/V_F Z_W$.

As was discussed earlier, Z_F can be approximated as LZ_O where L is the volume fraction lipid content of the fish, and Z_O is the Z value for octanol or lipid. K_B is then LK_{OW} where L is typically 0.05 or 5%.

From an examination of uptake data, Mackay and Hughes (1984) suggested that D_V is controlled by two resistances in series, a water resistance term D_W, and an organic resistance term D_O. Since the resistances are in series

$$1/D_V = 1/D_W + 1/D_O$$

The nature of the processes controlling D_W and D_O is not precisely known, but it is suspected that they are a combination of flow (GZ) and mass transfer (kAZ) resistances. If we substitute GZ for each D, recognizing that G may be fictitious, we obtain

$$1/k_2 = V_F Z_F / D_V = V_F L Z_O (1/G_W Z_W + 1/G_O Z_O)$$
$$= (V_F L/G_W)K_{OW} + (V_F L/G_O) \quad \text{(since } K_{OW} \text{ is } Z_O/Z_W)$$
$$= t_W K_{OW} + t_O$$

By plotting $1/k_2$ against K_{OW} for a series of chemicals taken up by gold-fish, Mackay and Hughes (1984) were able to estimate that t_W was about 0.001 hours and t_O 300 hours. This is another example of probing the nature of series or "two-film" resistances, using chemicals of different partition coefficient as discussed in Chapter 7.

The times t_W and t_O are believed to be characteristic of the fish species and vary with fish size and metabolic or respiration rate, as discussed by Gobas and Mackay (1987).

The uptake and clearance equilibria and kinetics, i.e., bioconcentration phenomena, of a conservative chemical in a fish are thus entirely described by K_{OW}, L, t_W and t_O.

8.7.3 Metabolism

If the fish can metabolize the chemical, the differential equation must be rewritten to include a metabolism term D_M, which is $V_F Z_F k_M$, where k_M is the metabolic rate constant. The differential equation then becomes

$$V_F Z_F df_F/dt = D_V(f_W - f_F) - D_M f_F$$

The rate constant k_2 is then increased to $(D_V + D_M)/V_F Z_F$, and the steady state fugacity of the fish will be reduced to $f_W D_V/(D_V + D_M)$, i.e., the fish will never achieve the water fugacity.

8.7.4 Growth Dilution

If the fish is growing, the assumption of constant V_F is invalid and the derivative becomes

$$d(V_F Z_F f_F)/dt = V_F Z_F df_F/dt + Z_F f_F dV_F/dt + V_F f_F dZ_F/dt$$

The last term can be ignored if the fish lipid content is constant. The middle term in dV_F is actually a Df group where

$$D_G = Z_F dV_F/dt$$

This is a "growth dilution" D value with units of mol/Pa.h, and reflects the reduction in fugacity caused by the same quantity of chemical being distributed, or diluted, in a larger volume of "solvent," the solvent here being fish. It is not a true process D value, but it is convenient to define it because it enables the differential equation to be rewritten as

$$V_F Z_F df_F/dt = D_V(f_W - f_F) - D_M f_F - D_G f_F$$

The rate constant is again increased, and a pseudo-steady state fugacity can be defined as

$$f_F = f_W D_V/(D_V + D_M + D_G)$$

The D_G term can become very important for hydrophobic chemicals for which the D_V and D_M terms are small. The primary determinant of concentration is then how fast the fish can grow and thus dilute the chemical. It should be noted that this treatment is simplistic in that as V_F increases the other D values also increase.

8.7.5 Uptake from Food (Biomagnification)

This process is more complicated and less understood than gill uptake. The first problem is quantifying the uptake efficiency, i.e., the ratio of quantity of chemical absorbed by the fish to chemical consumed. It is generally about 50% to 70%. Gobas et al. (1988b, 1989) have suggested that the uptake efficiency E_O from food in the gastrointestinal tract of a "clean" fish can also be described by a two-film approach yielding

$$1/E_O = A_W K_{OW} + A_O$$

where A_W and A_O are water and organic resistance terms similar in principle to t_W and t_O, but are dimensionless. A_O has a magnitude of about 2 and A_W a magnitude of about 10^{-7}, thus, for all but the most hydrophobic chemicals, E_O is about 50%. When K_{OW} exceeds 10^7, the efficiency falls off because of a high water phase resistance.

A D value D_A can be readily calculated for food consumption as $G_A Z_A$ where G_A is the food ingestion rate (m³/h), and Z_A is the Z value of the food, most readily estimated as $L_A Z_O$, where L_A is the lipid content of the food. Often the fugacity of a chemical in the food f_A will equal the fugacity in water. The rate of chemical uptake into the body of the fish is then $E_O.D_A.f_A$.

A major difficulty is encountered when describing the loss of chemical in feces and urine. In principle, a D value can be defined, but it is quite difficult and messy to measure G and Z; thus, neither are known. It is probable that the digestion process, which reduces both total mass and lipid content to provide matter and energy to the fish, reduces both G_A and Z_A, so that D_E for egestion is smaller than D_A. The simplest expedient is to postulate that it is reduced by a factor Q; thus, we can estimate D_E for transfer from the fish by egestion as $E_O D_A/Q$. Presumably the resistances causing E_O for uptake also apply to loss by egestion.

We can thus expand the differential equation to describe all processes

$$V_F Z_F df_F/dt = D_V(f_W - f_F) - D_M f_F - D_G f_F + E_O D_A f_A - E_O D_A f_F/Q$$

the rate constant k_2 is increased yet again to

$$(D_V + D_M + D_G + E_O D_A/Q)/V_F Z_F$$

and the steady state condition is

$$f_F = (D_V f_W + E_O D_A f_A)/(D_V + D_M + D_G + E_O D_A/Q)$$

Assuming that f_A equals f_W, i.e., food is in equilibrium with water, it is clear that f_F will approach f_W only when the D_V term dominates in both numerator and denominator. If K_{OW} is large, e.g. 10^6, the term D_A will tend to exceed D_V (because Z_A will greatly exceed Z_W) and the uptake of chemical in food becomes most important. The fish fugacity then tends toward Qf_A or Qf_W, i.e., the fish achieves a biomagnification factor of Q. Q is thus a maximum biomagnification factor, as well as being a D value ratio.

This biomagnification behavior has been observed by Connolly and Pedersen (1988), Q typically having a value of 3 to 5. Biomagnification is not immediately obvious until the fugacities are examined instead of the concentrations. At each step in the food chain, or at each trophic level there is a possibility of a fugacity multiple applying. It is thus apparent that fish fugacities and concentrations are a reflection of a complex combination of kinetic and equilibrium terms which can in principle be described by D values.

8.7.6 Food Chains

This simple model can be applied to a food chain starting with water then moving successively to phytoplankton, zooplankton, invertebrates, small fish, and to various levels of larger fish. Each level becomes food for the next higher level. If K_{OW} is relatively small, i.e., $< 10^5$, the D_V terms dominate and equifugacity is probable throughout the chain, i.e., f_A, f_W, and f_F will be equal. For larger K_{OW} chemicals, biomagnification is likely. For "superhydrophobic" chemicals of $K_{OW} > 10^7$ the E_O term becomes very small, uptake is slowed, and the growth and metabolism terms become critical. So, too, does association with suspended organic matter in the water column. A falloff in observed BCFs is (fortunately) observed for such chemicals; thus, there appears to be a "window" in K_{OW} about 10^6 to 10^7 in which bioaccumulation is most significant and most troublesome. DDT and PCBs lie in this "window." This issue has been discussed in detail and modeled by Thomann (1989).

Clark et al. (1988) have extended this thinking to include the fugacity of fish-eating birds, which are often most affected by hydrophobic chemicals. Several features of food chain biomagnification are of particular significance. Humans usually eat creatures close to the top of food chains, and strive to remain at the top of food chains, avoiding being eaten by other predators. Fish consumption is often the primary route of human exposure to hydrophobic chemicals. Creatures high in food chains are invaluable as bioindicators or biomonitors of contamination of lakes by hydrophobic chemicals. However, to use them as such requires knowledge of the D values, especially the D value for metabolism. A convincing argument can be made that if we live in an ecosystem in which wildlife at all trophic levels is thriving, we can be fairly optimistic that we humans are not being severely affected by environmental chemicals. This is thus a social incentive for developing, testing, and validating better environmental fate models, especially those employing fugacity.

8.7.7 Model

A computer model is provided on the diskette similar in format to those developed earlier. Regrettably, there are no generally validated correlations for the empirical terms t_O, t_W, A_O, A_W, and Q; thus, only illustrative values are given. The user is encouraged to determine real values from experimental data. It is expected that t_O and t_W will vary with fish size, probably increasing in proportion to fish size to a power such as 0.3, i.e., for two fish

$$t_1/t_2 = (V_1/V_2)^{0.3}$$

The terms A_O, A_W, and Q are probably not very dependent on fish size. The model includes a "bioavailability" calculation allowing for the water fugacity to be decreased by sorption to suspended matter in the water column. The model can be used to explore how variation in chemical properties changes bioaccumulation, and it can be run for a food chain by using fish concentration output from one calculation as food concentration input to the next.

8.8 INDOOR AIR MODELS

8.8.1 Introduction

We present here a very simple model of chemical fate in indoor air. Numerous studies have shown that humans are exposed to much higher concentrations of certain chemicals indoors than outdoors. Notable are radon, CO, CO_2, formaldehyde, pesticides, and volatile solvents present in glues, paints, and a variety of consumer products.

The key problem is that whereas advective flowrates are large outdoors, they are constrained to much smaller values indoors. Attempts to reduce heating costs often result in reduced air exchange, leading to increased chemical "entrapment." A nuclear submarine or a space vehicle is an extreme example of reduced advection. Fairly complicated models of chemical emission, sorption, reaction, and exhaust in multichamber buildings have been compiled, e.g., Nazaroff and Cass (1986, 1989), Thompson et al. (1986), but we treat here only a simple model developed by Mackay and Paterson (1983), which shows how D values can be used to estimate indoor concentrations caused by evaporating pools or spills of chemicals.

8.8.2 Model of Evaporating Chemical

We treat a situation in which a pool of chemical is evaporating into the basement air space of a two room (basement plus ground level) building with air circulation. If the building was entirely sealed and the chemical was nonreactive, evaporation would continue until the fugacity throughout the entire building equaled that of the pool (f_P). Of course, it is possible that the pool would have been completely evaporated by that time.

The evaporation rate can be characterized by a D value D_1 corresponding to kAZ, the mass transfer coefficient, pool area, air phase Z value product. If the chemical were in solution it would be necessary to invoke liquid and gas phase D values in series, i.e., the two film theory as discussed in Chapter 7.

The evaporated chemical may then be advected from one room to another with a D value D_2 defined as GZ, the air flow or exchange rate, air Z value product. From this second room it may be advected to the outdoors with another D value D_3. These advection rates are normally characterized as "air changes per hour" or ACH which is the advection rate G divided by the room or building volume and is the reciprocal of the air residence time. Typical ACHs for houses range from 0.25 to 1.5 per hour. The outdoor air has a defined background fugacity f_A.

It is apparent that the chemical experiences three D values in series in its journey from spill to outdoors; thus, the total D value will be given by

$$1/D_T = 1/D_1 + 1/D_2 + 1/D_3$$

and the flux N is $D_T(f_P - f_A)$ mol/h

Of interest are the intermediate fugacities in the rooms which can be estimated from the equations

$$N = D_1(f_P - f_1) = D_2(f_1 - f_2) = D_3(f_2 - f_A)$$

This is essentially a "three film" or "three resistance" model. Degrading reactions could be included leading to more complex, but still manageable, equations. Sorption to walls and floors could also be treated, but it is probably necessary to include these processes in the form of differential equations.

An example of a "spill" of a small quantity (1 g) of PCB over 0.01 m² (e.g., from a fluorescent ballast) was considered by Mackay and Paterson (1983). The PCB fugacity was 0.12 Pa and the outdoor concentration was taken as 4 ng/m³ or 3.7×10^{-8} Pa. The three D values (expressed as reciprocals) are

$$1/D_1 = 4900 \qquad 1/D_2 = 30 \qquad 1/D_3 = 15$$

D_T is thus essentially D_1, most resistance lying in the slow evaporation process from the small spill area. The evaporation rate N is then

$$N = D_T(f_P - f_A) = 2.4 \times 10^{-6} \text{ mol/h} = 6.4 \times 10^{-4} \text{ g/h}$$

The intermediate fugacities and concentrations are 3.6×10^{-5} Pa (3800 ng/m³) and 11×10^{-5} Pa (11500 ng/m³). The time for evaporation of 1 g of PCB will be 65 days. The amount of PCB in an air volume of 500 m³ would be of the order of 0.004 g, a small fraction of the small amount of PCB spilled.

The significant conclusion is that despite appreciable ventilation at a ACH of 0.5 h⁻¹, the indoor concentrations are over 1,000 times those outdoors. The indoor fugacity is, however, fortunately very much lower than the pool fugacity. Similar behavior applies to other solvents, pesticides, and chemicals which may be released indoors. Although the amounts spilled or released are small, the restricted advective dilution results in concentrations which are much higher than are normally encountered outdoors. In many cases this phenomenon is suspected to be the cause of the "sick building" problem in which residents complain repeatedly about headaches, nausea, and tiredness. The cure is to eliminate the source, or increase the ventilation rate. Fugacity calculations can contribute to understanding the problem.

No model is provided.

8.9 EXPOSURE, PHARMACOKINETICS, AND QSARs

8.9.1 Introduction

The calculations presented here lead to estimates of concentrations in air, water, soil, sediments, and fish. Perhaps the primary weakness is that they do not adequately address the problem of partitioning into vegetable matter which may be consumed by animals and humans. In this final section we discuss briefly the use of concentration data for assessment of impact on humans and other organisms.

The first obvious use of these concentrations is to compare them with concentration levels which are believed to cause adverse effects. These levels are usually developed by regulatory agencies and published as guidelines, objectives, or effect-concentrations of various types. The concentrations can also be used to estimate exposure or dosage, e.g., mg/day of chemical to an organism, which, for selfish reasons is usually a human. This calculation of dose is enlightening because it reveals which medium or route of exposure is of most importance. Presumably steps can be taken to reduce this route by, for example, restricting fish consumption.

This brings us to the issue of toxicity which we address here only briefly, being satisfied to point out the similarity between environmental partitioning calculations and pharmacokinetic models. It is also interesting to note the physical-chemical basis of quantitative structure activity relationships, or QSARs, which can be used to interpret, systematize, and even predict toxicity data.

8.9.2   Exposure

This material is taken essentially from a discussion of partitioning models by Mackay and Paterson (1988) in the text *Carcinogen Risk Assessment,* edited by Travis (1988).

Target or objective concentrations can be defined for most media. For example, from considerations of toxicity or aesthetics it may be possible to suggest that water concentrations should be maintained below 1 mg/m^3, air below 1 μg/m^3, and fish below 1 mg/kg. These concentrations can be compared as a ratio or quotient to the estimated environmental concentrations. A hypothetical example is given in Table 8.3 illustrating the quotient method. It is apparent that the primary concern is with air inhalation and fish ingestion. The proximities of the estimated prevailing concentrations to the targets are expressed as these quotients, a large value implying a large safety factor and low risk. The high-risk situations correspond to low quotients. The concentration level in fish may not be directly toxic to fish, but may pose a threat to humans if the fish is consumed on a regular basis.

Difficulties are encountered when suggesting target concentrations in soil and sediment, because these media are not normally consumed directly by organisms of commercial interest. Whereas simple lethality experiments can be designed using air, water, or food as vehicles of toxicant administration, it

Table 8.3. Environmental Concentrations, Human Intake Rates and Effect Levels.

Comparison of Environmental Effect Concentrations with Predicted Concentrations

	Effect Level	Actual Level	Quotient or Ratio
Air pollution (mg/m³)	15	2.82	5.3
Fish toxicity (g/m³)	5	2.91×10^{-4}	17,160
Soil contamination (µg/g)	5	3.58×10^{-2}	140

Comparison of Allowable Human Intake with Predicted Intakes[a]

	Concentration (g/m³)	Volume of Intake (m³/day)	Actual Intake (µg/day)	Percent of Permissible Intake	Percent of Actual Intake
Ingestion of food and water					
Fish	0.140	0.00025	35	0.7	34
Meat products	0.140	0.00005	7	0.14	6.8
Dairy products	2.9×10^{-4}	0.00025	0.073	0.001	0.07
Vegetable products	2.9×10^{-4}	0.00045	0.13	0.003	0.13
Water	2.9×10^{-4}	0.002	0.58	0.012	0.56
Air inhalation	2.82×10^{-6}	20	56	1.1	54.4
Domestic exposure			1	0.02	1
Occupational exposure			3	0.06	3
Total			103[b]	2.04	100

[a]Allowable daily intake (ADI) = 5000 µg/day.
[b]Ratio of ADI/intake = 5000/103 = 48.5.

is not always clear how concentrations in the solid matrices of soils and sediments relate to exposure or intake of chemical by organisms. It is difficult to design meaningful bioassays involving interactions between organisms, soils, and sediments. One approach is to decree that whatever target fugacity is developed for water, be applied to sediment. This effectively links the target concentrations by equilibrium partition coefficients.

The dosage approach uses these media concentrations to calculate dosages to human and other organisms, as illustrated in Table 8.3 and Figure 8.8.

An average human inhales some 20 m^3 of air per day; thus, the amount of chemical inhaled in this air can be easily calculated in units of μg/day. Not all this chemical may be absorbed, but at least a maximum dosage can be deduced. The same human may consume 2 liters/day of water containing dissolved chemical, enabling this dosage to be estimated again in μg/day. Food, the other vehicle, is more difficult to estimate. A typical diet may consist of 1 liter/day of solids broken down as shown in Table 8.3. Fish concentrations can be estimated directly from water concentrations, but meat, vegetable, and dairy product concentrations are still poorly understood functions of the concentrations of chemical in air, water, soil, animal feeds, and of agrochemical usage. Techniques are emerging for calculating food-environment concentration ratios, but at present the best approach is to analyze a typical purchased "food basket." The food concentrations used in Table 8.3 are illustrative and were obtained by assuming that meat has partitioning properties similar to fish, and that dairy and vegetable products consist largely of water containing some fat or lipids. This issue is complicated by the fact that much food is grown at distant locations and imported. Beverages, food, and water may also be treated for chemical removal commercially or domestically by washing, peeling, or cooking.

Significant chemical exposure may also occur in the occupational setting (e.g., factories) in institutional and commercial facilities (e.g., schools, stores, and cinemas) and at home, but these exposures vary greatly from individual to individual and depend on lifestyle. For illustrative purposes and to give a complete picture, exposures from these sources are included in Table 8.3 and pictorially in Figure 8.8.

There emerges a profile of relative exposures by various routes from which the dominant route(s) can be identified. If desired, appropriate measures can be taken to reduce the largest exposures. The advantage of this approach is that it places the spectrum of exposure routes in perspective. There is little merit in striving to reduce an already small exposure.

The dominant routes in the case treated in Table 8.3 are consumption of fish and air inhalation. Routes vary greatly in magnitude and relative

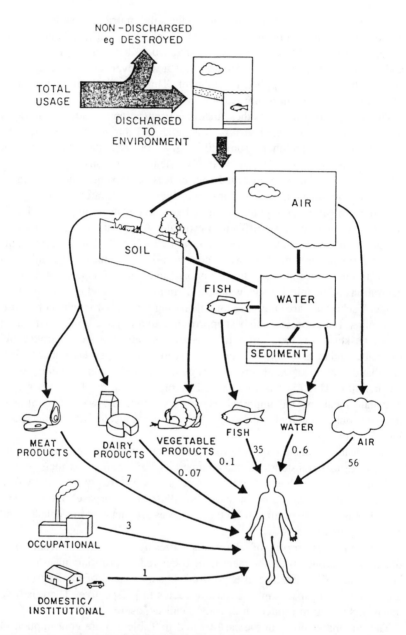

Figure 8.8.   Illustration of routes of human exposure of a hypothetical compound (μg/day).

contribution, depending on the nature of the chemical and especially its physical chemical properties.

Data from the environmental fate models can provide a sound basis for estimating risk when used to assess quotients, and determine dominant exposure routes. If such information can be presented to the public the individual is, at least in principle, able to choose or modify a lifestyle to minimize exposure and presumably risk. Individuals then have the freedom and information to judge and respond to acceptability of risk from exposure to this chemical, compared to the other voluntary and involuntary risks to which they are subject.

8.9.3 QSARs and Pharmacokinetics

Much of the concern about environmental contamination by chemicals relates to their toxic effects on organisms, especially humans. An elementary principle of toxicology is that it is the presence and quantity or dose of chemical at some target site in the body, e.g., the liver, which causes the toxic effect. The dose makes the poison. The external concentration in air, water, or food is only an indirect indication of the concentration at the target site, but it is these external concentrations which are most easily measured and regulated, and it is these concentrations which are calculated in this book. There is thus an incentive to continue the chemical fate calculation from the environment into the body of the victim, and estimate the concentration at the target. A chemical's toxicity is usually expressed as an external concentration, e.g., 100 mg/m^3 in air, which causes a defined toxic effect, the most obvious and final being death. If two chemicals have different toxicities of, for example, 1 and 100 mg/m^3 in air, is it possible that they are really equally toxic because their partitioning properties are such that both these concentrations result in a single concentration of 500 μg/g in the liver? If this is so, the interpretation of toxicity information and its extrapolation from chemical to chemical becomes easier. Prediction of toxicity may even be possible, especially for homologous series of chemicals.

This idea is best illustrated for the well-studied and convenient system of fish toxicity measurement or bioassay involving results from exposure of the fish to chemical in solution in water. It is found that death or loss of consciousness (narcosis) occurs after exposure to defined concentrations for a defined period of time, such as 48 or 96 hours. The concentration causing death to 50% of a group of fish is usually referred to as the lethal concentration (LC) to 50% of the fish after 48 hours or a "48 hour LC50." It is usual to obtain data for a variety of chemicals and plot the LC50 versus various molecular descriptors or properties such as carbon number, molecular mass or volume, solubility, or, most popular, the octanol water partition coefficient. Quantitative relationships between structure and activity or QSARs are sought and established. Most celebrated of these relationships are those of Veith et al. (1983), Konemann (1981) and the many studies by Schultz and Hermans which are reviewed in the compilations by Kaiser (1984, 1987) and Karcher and Devillers (1990). The general form of the correlations is exemplified by that of Konemann for guppies that

$$\log (1/LC50) = 0.87 \log K_{OW} - 4.87$$

where LC50 has units of $\mu mol/L$. A low LC50 implies a high toxicity which accounts for the use of the reciprocal, i.e., (1/LC50) as a measure of toxicity. It thus appears that as K_{OW} increases by, for example, adding chlorines or methyl groups, toxicity increases because the concentration necessary to cause the effect decreases. For example, chemicals with log K_{OW} values of 3 and 5 will have LC50s of 182 and 3.3 $\mu mol/L$, a factor of 55 difference.

If we assume that the target site at which the toxicity is exerted is similar to octanol in its partitioning properties, that the chemical reaches equilibrium during the test, and that it is not appreciably metabolized, then these concentrations of 182 and 3.3 $\mu mol/L$ will result in concentrations at the target of 182 K_{OW} or 182000 $\mu mol/L$ and 3.3 K_{OW} or 330000 $\mu mol/L$ at the target site, which differ by only a factor of 1.8. Indeed the chemical with log K_{OW} of 3 now appears to be more toxic because it causes the effect at a lower target site concentration. If the coefficient on log Kow was 1.0 instead of 0.87 the target site concentrations would be equal. Significantly, the coefficient is usually close to unity.

These concentrations correspond, for molecules of molar volume 200 cm^3/mol, to remarkably large concentrations of 36 and 66 cm^3/L or volume fractions of 0.036 and 0.066. The whole body concentrations of these chemicals in a fish of 5% lipids are probably about a factor of 20 lower, i.e., about 10 mmol/L, or a volume fraction of 0.002. To a first approximation these narcotic chemicals have fairly similar concentrations at the target site and appear to disrupt some critical process by virtue of occupying such a large volume fraction of some essential tissue. They may "swell" the site to such an extent that it no longer functions properly. The narcosis is reversible, i.e., on exposure to clean water the fish will clear the chemical and regain consciousness.

The key conclusion is that by examining the fish-water partitioning phenomena it is possible to interpret the external LC50 data in terms of a fairly constant chemical concentration at some internal target site which appears to have properties similar to octanol. This approach has been more fully discussed by Abernethy et al. (1988). The predictive feature derived from these relationships is invaluable as a means of reducing the need for widespread toxicity testing.

For mammals, especially humans, the situation is more complex because it is not as easy to manipulate concentrations in the respiring medium of air. Administration is often oral. The times for equilibration may be longer and there are often more sophisticated enzyme systems working to metabolize the chemicals. Most promising as a sophisticated "substitute" for QSARs are physiologically based pharmacokinetic models (PBPK models) which treat uptake from air, food, and possibly by dermal contact or injection and calculate the nature of circulation of the chemical in venous and arterial blood, to and from various organs or tissue groups including adipose tissue, muscle, skin, brain, kidney, and liver. As in environmental models, partition coefficients or Z values can be deduced for chemical equilibration between air, blood, and various organs. Flows of blood to each organ can be expressed as D values. Metabolism rates can be expressed using rate constants, usually invoking Michaelis-Menten kinetics and translated into D values. Mass balance equations can then be assembled describing steady and unsteady state conditions following exposure to pulses or long-term constant concentrations.

Much of the pharmacokinetic literature is devoted to assessment of the time course of the fate of therapeutic drugs within the human body, the aim being to supply a sufficient, but not too large and thus toxic, dose of drug to the target organ. Closer to environmental exposure conditions are PBPK models for occupational exposure to toxicants such as solvent or fuel vapors which may be intermittent or continuous in nature. An example is the model of Ramsey and Andersen (1984) which was translated into fugacity terms by Paterson and Mackay (1986, 1987). Accounts of various aspects of pharmacokinetics and PBPK models and their contribution to environmental science are the works of Welling (1986), Parke (1982), Reitz and Gehring (1982), Tuey and Matthews (1980), Fiserova-Bergerova (1983), and Menzel (1987).

Figure 8.9 from Paterson and Mackay (1987) illustrates the fugacity approach to modeling the fate of a chemical in the human. In principle, it is possible to calculate steady and unsteady state fugacities, concentrations, amounts, fluxes, and response times, thus linking external environmental concentrations to internal tissue concentrations. Ultimately from a human health viewpoint it is likely that it will be possible to undertake these calculations and compare levels of chemical contamination in vulnerable tissues with levels which are known to cause adverse effects. Most difficult is likely to be the addition of the toxic effects of numerous chemicals to give a total effect; the treatment of chemicals which are not in themselves very toxic but are metabolically converted into more active forms or which adversely affect the immune system; and the assessment of risk to the organism of genotoxicity such as cancer which may be induced by low concentrations of chemical after a prolonged latent period, and profoundly affected by the failure of biochemical repair systems.

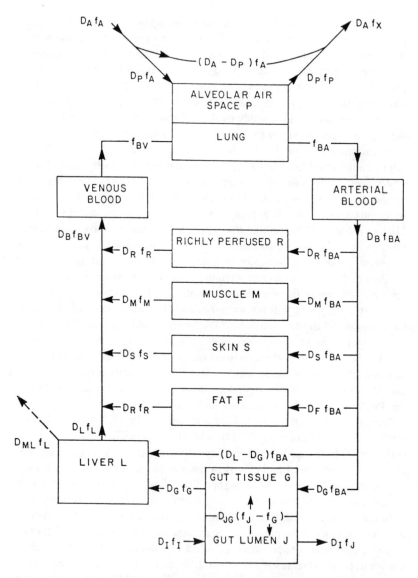

Figure 8.9.   The fugacity approach to pharmacokinetic models.

There is clearly a need to link the environmental and pharmacokinetic modeling efforts to build up a comprehensive capability of assessing the journey of the chemical from source to environment, to organism, and ultimately to the target site.

8.10 CLOSURE

Perhaps the task addressed by this book is best summarized by Figure 8.10, which depicts many of the environmental processes to which chemical contaminants are subject. The aim has been to develop methods of calculating partitioning, transport, and transformation in the wide range of media which comprise our environment. Ultimately of primary concern to the public, and thus to regulators, is the effect which these chemicals may have on human well-being. But, as was discussed earlier, there are sound practical and ethical reasons for protecting wildlife, and indeed all our fellow organisms which inhabit our ecosystem.

It is not yet clear how severe the effects of chemical contaminants are, nor is it likely that the full picture will become clear for some decades. Undoubtedly there are chemical surprises or "time bombs" in store as analytical methods and toxicology improve.

Regardless of the incentive nurtured by public fear of "toxics," environmental science has a quite independent and noble objective of seeking, for its own sake, a fuller quantitative understanding of how the biotic and abiotic components of our multimedia ecosystem operate, how chemicals which enter this system are transported, transformed, and accumulate, and how they eventually reach organisms and affect their well-being. The prudent course of action is for society to reduce discharges to a level at which there is assurance that there are no adverse effects of chemicals on the quality of life, singly or in combination. It is hoped that the tools developed in these chapters can contribute to this process.

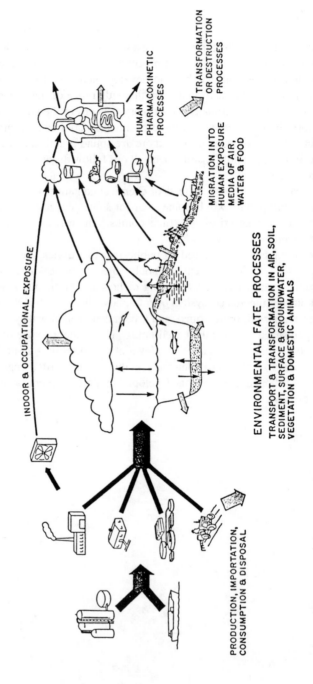

Figure 8.10. An illustration of a chemical's sources, environmental fate, human exposure, and human pharmacokinetics.

Appendix 1   FUGACITY FORMS

FUGACITY FORM 1	Z VALUES

CHEMICAL:

Temperature: _____ °C   _____ K $RT = 8.314 \times$ _____ =

Molecular mass: _____ g/mol (w)

Water solubility: _____ g/m^3 or mg/L _____ mol/m^3 (C^S)

Vapor pressure: _____ Pa (P^S) _____ mm Hg _____ atm

Log K_{OW}: _____ K_{OW} _____

AIR-WATER PARTITION COEFFICIENTS AND Z VALUES

$H = P^S/C^S =$ _____ Pa.m^3/mol

$K_{AW} = H/RT =$ _____

Z_A (air) = 1/RT = _____   Air density = 0.029 x 101325/RT

Z_W (water) = 1/H = _____   = _____ kg/m^3

OTHER PHASES

Name	
Density ρ kg/m^3	
Organic carbon or lipid content φ g/g	
K_{OC} or K_{OL} [1]	
$K_P = \phi K_{OC}$ or ϕK_{OL}	
$K_{PW} = K_P\rho/1000$	
$Z_P = K_{PW} \cdot Z_W$	

(1) e.g. $K_{OC} = 0.41\ K_{OW}$, $K_{OL} = K_{OW}$

FUGACITY FORM 2		LEVEL I
CHEMICAL:	Amount M mol	kg
Compartment		
Volume V m^3		
Z mol/m^3 Pa		
VZ mol/Pa		
	$\dfrac{\text{Amount M mol}}{\Sigma VZ} = \underline{\hspace{2cm}} =$ $= $ FUGACITY f	
C = Zf mol/m^3 m = CV mol percent		
C_G g/m^3 [(1)] Density ρ kg/m^3 C_U μg/g [(2)]		

(1) C_G = C x Molecular Mass (g/mol)
(2) C_U = C_G x 1000/Density (kg/m^3)

FUGACITY FORM 3   LEVEL II

CHEMICAL:

Direct emission rate E _____ mol/h

Advective input rates

Compartment

Volume m^3 (V)

Residence time h (t)

Flow rate m^3/h = V/t = G

Inflow concentration mol/m^3 C_B

Chemical inflow rate mol/h = GC_B

Total input rate E + ΣGC_B = I =

Compartment

Volume V m^3

Z

VZ

Reaction half life (h) t

Rate constant k = 0.693/t (h^{-1})

Advective flow G m^3/h

D reaction = VZk = D_R

D advection = GZ = D_A

$D_R + D_A = D_T$

Total D value = ΣD_T =   Fugacity f = I/ΣD =

C = Zf mol/m^3

m = CV mol

percent

C_G g/m^3, i.e. CW

Density ρ kg/m^3

C_U $\mu g/g$, i.e. $C_G \times 1000/\rho$

Reaction rate $D_R f$

Advection rate $D_A f$

Total rate $D_T f$

Total amount M = Σm =

Total reaction rate = $\Sigma D_R f$ =   Reaction residence time (h) = M/$\Sigma D_R f$ =

Total advection rate = $\Sigma D_A f$ =   Advection residence time (h) = M/$\Sigma D_A f$ =

Total output rate (mol/h) = I =   Overall residence time (h) = M/I =

Appendix 2   COMPUTER PROGRAMS

The diskette contains 11 programs written in BASIC language and saved in BASIC. They can be retrieved, run, and modified using one of the BASIC programs such as IBM Microsoft or GW BASIC. After conversion to ASCII code, they can be used on other versions such as TurboBasic.

It is suggested that the user make a copy and write-protect the diskette.

The programs are fully commented to aid interpretation. The nomenclature is similar, but not always identical to that used in the text. Some programs require condensed print to accommodate the line width; thus the instruction to the printer may require modification.

"LEVEL1A" is a Level I program treating four compartments. It prompts for chemical properties and amount. The phase volumes and properties can only be changed by editing the program.

"LEVEL1B" is a six-compartment Level I program, similar to LEVEL1A.

"LEVEL2A" is a four-compartment Level II program which prompts for the same information as Level I programs, but also for reaction and advection data.

"LEVEL2B" is similar, but treats six compartments.

"LEVEL3A" is a four-compartment Level III program which requires all the Level II data and prompts for D values in the form of transfer half-lives.

"LEVEL3B" is similar to LEVEL3A, but prompts for the D values directly.

"AIRWATER" treats air-water exchange processes as described in Chapter 8.2. It, and the other vignette models, contain data for at least one chemical, but can prompt for data for other chemicals, and can rerun the program for the previously used chemical; i.e., the program returns to the start. Similarly, illustrative, new, or previously specified environmental data can be entered.

"SOIL" treats chemical behavior in a surface soil as described in Chapter 8.3.

"SEDIMENT" treats sediment-water exchange as outlined in Chapter 8.4.

"FISH" treats fish bioaccumulation from water and food as discussed in Chapter 8.7.

"QWASI" is a steady-state model of chemical fate in a water-air sediment system as presented in Chapter 8.5. This model is longer and more complex than the others and must be edited to treat a specific lake system.

The user is encouraged to view these as only skeletons on which more complex and site-specific models can be constructed.

REFERENCES AND BIBLIOGRAPHY

Abernethy, S., Bobra, A. M., Shiu, W. Y., Wells, P. G., and Mackay, D. (1986) Acute lethal toxicity of hydrocarbons and chlorinated hydrocarbons to two planktonic crustaceans: The key role of organism-water partitioning. *Aquat. Toxicol.* 8:163–174.

Abernethy, S., Mackay, D., and McCarthy, L. S. (1988) Volume fraction correlation for narcosis in aquatic organsims: The key role of partitioning. *Environ. Toxicol. Chem.* 7:469–481.

Alexander, M. (1981) Biodegradation of chemicals of environmental concern. *Science* 211:132–138.

Alexander, M. (1985) Biodegradation of organic chemicals. *Environ. Sci. Technol.* 18:106–111.

Ashworth, R. A., Howe, G. B., Mullins, M. E., and Rogers, T. N. (1988) Air-water partitioning coefficients of organics in dilute aqueous solutions. *J. Hazardous Materials* 18:25–36.

Bacci, E., Calamari, D., Gaggi, C., and Vighi, M. (1990) Bioconcentration of organic chemical vapors in plant leaves: experimental measurements and correlation. *Environ. Sci. Technol.* 24:885–889.

Banerjee, S., Yalkowsky, S. H., and Valvani, S. C. (1980) Water solubility and octanol/water partition coefficients of organics. Limitations of the solubility-partition coefficient correlation. *Environ. Sci. Technol.* 14:1227–1229.

Banerjee, S., Howard, P. H., Rosenberg, A. M., Dombrowski, A. E., Sikka, H., and Tullis, D. L. (1984) Development of a general kinetic model for biodegradation and its application to chlorophenols and related compounds. *Environ. Sci. Technol.* 18:416–422.

Baughman, G. L., and Lassiter, R. R. (1978) *in* Estimating the hazard of chemical substances to aquatic life. J. Cairns, Jr., K. G. Dickson, A. W. Maki, eds., American Society of Testing and Materials Tech. Pub. 657, Philadelphia, PA.

236   MULTIMEDIA ENVIRONMENTAL MODELS AND FUGACITY

Bidleman, T. F. (1984) Estimation of vapor pressures for nonpolar organic compounds by capillary gas chromatography. *Anal. Chem.* 56:2490–2495.

Bidleman, T. F., Billings, W. N., and Foreman, W. T. (1986) Vapor-particle partitioning of semivolatile organic compounds: Estimates from field collections. *Environ. Sci. Technol.* 20:1038–1042.

Bidleman, T. F. (1988) Atmospheric processes. *Environ. Sci. Technol.* 22:361–367 and errata 726.

Bird, R. B., Stewart, W. E., and Lightfoot, E. N. (1960) *Transport Phenomena.* John Wiley & Sons, New York, NY.

Boethling, R. S., and Sabijic, A. (1989) Screening-level model for aerobic biodegradability based on a survey of expert knowledge. *Environ. Sci. Technol.* 23:672–679.

Branson, D. R., Blau, G. E., Alexander, H. C., and Neely, W. B. (1975) Bioconcentration of 2,2′,4,4′-tetrachlorobiphenyl in rainbow trout as measured by an accelerated test. *Trans. Am. Fish Soc.* 104:785–792.

Brown, Jr., J. F., Bedard, D. L., Brennan, M. J., Carnahan, J. C., Feng, H., and Wagner, R. E. (1987) Polychlorinated biphenyl dechlorination in aquatic sediments. *Science* 236:709–712.

Brutsaert, W., and Jirka, G. H. (Eds.) (1984) *Gas Transfer at Water Surfaces.* D. Reidel Publ. Co., Dordrecht, Holland.

Burkhard, L. P., Andren, A. W., and Armstrong, D. E. (1985) Estimation of vapor pressures for polychlorinated biphenyls: A comparison of eleven predictive methods. *Environ. Sci. Technol.* 19:500–506.

Butkovic, V., Klasinc, L., Orhanovic, M., Turk, J., and Gusten, H. (1983) Reaction rates of polynuclear aromatic hydrocarbons with ozone in water. *Environ. Sci. Technol.* 17:546–547.

Calamari, D., Vighi, M., and Bacci, E. (1987) The use of terrestrial plant biomass as a parameter in the fugacity model. *Chemosphere* 16:2359–2364.

Callahan, M. A., Slimak, M. W., Gabel, N. W., May, I. P., Fowler, C. F., Freed, J. R., Jennings, P., Durfee, R. L., Whitmore, F. C., Maestri, B., Mabey, W. R., Holt, B. R., and Gould, C. (1979) Water-Related Environmental Fate of 129 Priority Pollutants, Vols. I and II, EPA Report No. 440/4-79-029a. Versar, Inc., Springfield, VA.

Canadian Environmental Advisory Council (CEAC) (1988) Listing toxics under CEPA—Is the chemistry right? Environment Canada, Ottawa.

Carson, R. (1962) *Silent Spring,* Houghton Mifflin, Boston, MA.

Clark, T., Clark, K., Paterson, S., Nostrom, R., and Mackay, D. (1988) Wildlife monitoring, modelling and fugacity. *Environ. Sci. Technol.* 22:120–127.

Clark, K. E., Gobas, F. A. P. C., and Mackay, D. (1990) Model of organic chemical uptake and clearance by fish from food and water. *Environ. Sci. Technol.* 24:1203-1213.

Cohen, Y., and Ryan, P. A. (1985) Multimedia modeling of environmental transport: Trichloroethylene test case. *Environ. Sci. Technol.* 9:412.

Cohen, Y., Ed. (1986) *Pollutants in a Multimedia Environment.* Plenum Press, N.Y.

Connolly, J. P., and Pedersen, C. J. (1988) A thermodynamically based evaluation of organic chemical accumulation in aquatic organisms. *Environ. Sci. Technol.* 22:99-103.

Conway, R. A. (1982) *Environmental Risk Analysis for Chemicals.* Van Nostrand Reinhold, New York, NY.

Csanady, G. T. (1973) *Turbulent Diffusion in the Environment.* D. Reidel Publ. Co., Dordrecht, Holland.

Dean, J. D., Ed. (1985) *Lange's Handbook of Chemistry.* 13th ed., McGraw-Hill, New York, NY.

Dearden, J. C. (1990) Physico-chemical descriptors, in practical applications of quantitative structure-activity relationships (QSAR) in environmental chemistry and toxicology, Karcher, W. and Devillers, J., Eds. Kluwer Academic Publisher, Dordrecht, Holland. 25-60.

Denbigh, K. G. (1966) *The Principles of Chemical Equilibrium,* 2nd Ed., Cambridge University Press, London.

Dilling, W. L., Bredeweg, C. J., and Tefertiller, N. B. (1976) Organic Photochemistry. Simulated atmospheric photocomposition rates of methylene chloride, 1,1,1-trichloroethane, tetrachloroethylene, and other compounds. *Environ. Sci. Technol.* 10:351-356.

DiToro, D. M. (1985) A particle interaction model of reversible organic chemical sorption. *Chemosphere* 14:1503-1538.

Dobbs, A. J., and Cull, M. R. (1982) Volatilization of chemicals—relative loss rates and the estimation of vapor pressures. *Environ. Pollut.* 3:289-298.

Duinker, J. C., and Bouchertall, F. (1989) On the distribution of atmospheric polychlorinated biphenyl congeners between vapor phase, aerosols, and rain. *Environ. Sci. Technol.* 23:57-62.

Dulin, D., and Mill, T. (1982) Development and evaluation of sunlight actinometers. *Environ. Sci. Technol.* 16:815-820.

Dunn, W. J., Block, J. H., and Pearlmonk, S. (1986) *Partition Coefficient: Determination & Estimation.* Pergamon Press, Elmsford, N.Y.

Eastcott, L., Shiu, W. Y., and Mackay, D. (1988) Environmentally relevant physical chemical properties of hydrocarbons: A review of data and development and simple correlations. *Oil and Chemical Pollution* 4:191-216.

Eisenreich, S. J. (1987) The chemical limnology of nonpolar organic contaminants. PCBs in Lake Superior. Ch. 13, pp. 393–470, in *Sources and Fates of Aquatic Pollutants*. Hites, R. A., and Eisenreich, S. J., Eds. American Chemical Society, Adv. in Chem. Series 216, Washington, DC.

EPA (1982) U.S. Environmental Protection Agency Report "Chemical Fate Test Guidelines." No. 56016-82-003. Washington, DC.

Faust, B. C., and Holgne, J. (1987) Sensitized photooxidation of phenols by fulvic acid and in natural waters. *Environ. Sci. Technol.* 21:957–1064.

Fendinger, N. J., and Glotfelty, D. E. (1988) A laboratory method for the experimental determination of air-water Henry's law constants for several pesticides. *Environ. Sci. Technol.* 22:1289–1293.

Fiserova-Bergerova, V. (1983) *Modeling of Inhalation Exposure to Vapors*, Vols. 1 and 2. CRC Press, Boca Raton, FL.

Foreman, W. T., and Bidleman, T. F. (1985) Vapor pressure estimates of individual polychlorinated biphenyls and commercial fluids using gas chromatographic retention data. *J. Chromatog.* 330:203–216.

Foreman, W. T., and Bidleman, T. F. (1987) An experimental system for investigating vapor-particle partitioning of trace organic pollutants. *Environ. Sci. Technol.* 21:869–875.

Formica, S. J., Baron, J. A., Thibodeaux, L. J., and Valsaraj, T. (1988) PCB transport into lake sediments. Conceptual model and laboratory simulation. *Environ. Sci. Technol.* 22:1435–1440.

Friesen, K. J., Vilk, J., and Muir, D. C. G. (1990) Aqueous solubilities of selected 2,3,7,8—substituted polychlorinated dibenzofurans (PCDFs). *Chemosphere* 20:27–32.

Gauthier, T. D., Seltz, W. R., and Grant, C. L. (1987) Effects of structural and compositional variations of dissolved humic materials on pyrene K_{OC} values. *Environ. Sci. Technol.* 21(3):243–248.

Ghosal, D., You, I. S., Chatterjee, D. K., and Chakrabarty, A. M. (1985) Microbial degradation of halogenated compounds. *Science* 228:135–142.

Glotfelty, D. E., Seiber, J. N., and Liljedahl, L. A. (1987) Pesticides in fog. *Nature* 325:602–608.

Gobas, F. A. P. C., and Mackay, D. (1987) Dynamics of hydrophobic organic chemical bioconcentration in fish. *Environ. Toxicol. Chem.* 6:495–504.

Gobas, F. A. P. C., Lahitette, J. M., Garofalo, G., and Mackay, D. (1988a) Membrane-water versus 1-octanol-water partitioning: A novel method to determine membrane-water partition coefficients. *J. Pharm. Sci.* 77:265–272.

Gobas, F. A. P. C., Muir, D. C. G., and Mackay, D. (1988b) Dynamics of dietary bioaccumulation and faecal elimination of hydrophobic organic chemicals in fish. *Chemosphere* 17:943–962.

Gobas, F. A. P. C., Clark, K. E., Shiu, W. Y., and Mackay, D. (1989) Bioconcentration of polybrominated benzenes and biphenyls and related superhydrophobic chemicals in fish: Role of bioavailability and elimination into the feces. *Environ. Toxicol. Chem.* 8:231–245.

Gossett, J. M. (1987) Measurement of Henry's law constants for C_1 and C_2 chlorinated hydrocarbons. *Environ. Sci. Technol.* 21:202–208.

Green, J. C. (1988) Classification and properties of soils. Appendix C in Book I, Lyman, W. J., Reehl, W. F. and Rosenblatt, D. H., Eds. *Environmental Inorganic Chemistry: Properties Processes and Estimation Methods.* Pergamon Press, New York, NY.

Hansch, D., and Leo, A. (1979) *Substituent Constants for Correlation Analysis in Chemistry and Biology.* Wiley-Interscience, New York, NY.

Hansen, P. C., and Eckert, C. A. (1986) An improved transpiration method for the measurement of very low vapor pressures. *J. Chem. Eng.* 31:1–3.

Hites, R. A., and Eisenreich, S. J. (1987) *Sources and Fates of Aquatic Pollutants.* Adv. in Chem. Series 216, American Chemical Society, Washington, DC.

Holysh, M., Paterson, S., Mackay, D., and Bandurraga, M. M. (1985) Assessment of the environmental fate of linear alkylbenzenesulphonates. *Chemosphere* 15:3–20.

Horvath, A. L. (1982) *Halogenated Hydrocarbons, Solubility-Miscibility with Water.* Marcel Dekker, Inc., New York, NY.

Horvath, A. L., and Getzen, F. W. (1985) IUPAC Solubility Data Series, Vol. 20. *Halogenated Benzenes, Toluenes, and Phenols with Water.* Pergamon Press, Oxford, UK.

Howard, P. H., and Banerjee, S. (1984) Interpreting results from biodegradability tests of chemicals in water and soil. *Environ. Toxicol. Chem.* 3:551–562.

Howard, P. H., Heuber, A. E., Mulesky, B. C., Crisman, J. S., Meylan, W., Crosbie, E., Gray, D. A., Sage, G. W., Howard, K. P., and LaMacchia, A. (1986) Biolog, biodeg, and fate/expos: New files on microbial degradation and toxicity as well as environmental fate/exposure of chemicals. *Environ. Toxicol. Chem.* 5:977–988.

Howard, P. H., Hueber, A. E., and Boethling, R. S. (1987) Biodegradation data evaluation for structure/biodegradability relations. *Environ. Toxicol. Chem.* 6:1–10.

Howard, P. H. (1989) *Handbook of Environmental Fate and Exposure Data for Organic Chemicals,* (several volumes) *Large Production and Priority Pollutants,* Vol. I. Lewis Publishers, Chelsea, MI.

Hutzinger, O., Ed. *The Handbooks of Environmental Chemistry* (various editions and years), Springer-Verlag, Berlin.

IUPAC (1985) *Solubility Data Series Vol. 20: Halogenated Benzenes, Toluenes and Phenols with Water.* Horvath, A. L., and Getzen, F. W., Eds., Pergamon Press, Oxford.

IUPAC (1989) *Solubility Data Series, Vols. 37 and 38, Hydrocarbons with Water and Seawater, Part I. Hydrocarbons C_5 to C_7; Part II. Hydrocarbons C_8 to C_{36}.* Shaw, D., Ed., Pergamon Press, Oxford.

Jeffers, P. M., Ward, L. M., Woytowitch, L. M., and Wolfe, N. L. (1989) Homogenous hydrolysis rate constants for selected chlorinated methanes, ethanes, ethenes, and propanes. *Environ. Sci. Technol.* 23:968.

Johnson, B. T. (1980) *Evaluation of Methods. Approaches to Estimating Microbial Degradation of Chemical Contaminants in Freshwater. Biotransformations and Fate of Chemicals in the Environment.* Maki, A. W., Dickson, K. L., and Cairns, J., Eds. American Society of Microbiology, Washington, DC, pp. 25–33.

Jorgensen, S. E., Ed. (1984) *Modelling the Fate and Effect of Toxic Substances in the Environment.* Elsevier, Amsterdam.

Jorgensen, S. E., and Gramiec, M. J., Eds. (1989) *Mathematical Submodels in Water Quality Analysis.* Elsevier, Amsterdam.

Jury, W. A., Spencer, W. F., and Farmer, W. J. (1983) Behavior assessment model for trace organics in soil: I Model description. *J. Environ. Qual.* 12:558.

Jury, W. A., Farmer, W. J., and Spencer, W. F. (1984) Behavior assessment model for trace organics in soil: II Chemical classification and parameter sensitivity. *J. Environ. Qual.* 13:567.

Jury, W. A., Farmer, W. J., and Spencer, W. F. (1984) Behavior assessment model for trace organics in soil: III Application of screening model. *J. Environ. Qual.* 13:573.

Jury, W. A., Spencer, W. F., and Farmer, W. J. (1984) Behavior assessment model for trace organics in soil: IV Review of experimental evidence. *J. Environ. Qual.* 13:580.

Kaiser, K. L. E. (1984) QSAR in Environmental Toxicology. D. Reidel Publishing Co., Dordrecht, Holland.

Kaiser, K. L. E. (1987) QSAR in Environmental Toxicology II. D. Reidel Publishing Co., Dordrecht, Holland.

Kamlet, M. J., Doherty, R. M., Abboud, J. L., Abraham, M. H., and Taft, R. W. (1986) Solubility: A new look. *Chemtech.* Sept.:566–576.

Kamlet, M. J., Doherty, R. M., Carr, P. W., Mackay, D., Abraham, M. H., and Taft, R. W. (1988) Linear solvation energy relationships. 44.

Parameter estimation rules that allow accurate prediction of octanol/water partition coefficients and other solubility and toxicity properties of polychlorinated biphenyls and polycyclic aromatic hydrocarbons. *Environ. Sci. Technol.* 22:503–509.

Karickhoff, S. W. (1981) Semiempirical estimation of sorption of hydrophobic pollutants on natural sediments and soils. *Chemosphere* 10:833–849.

Karickhoff, S. W. (1984) Organic pollutant sorption in aquatic systems. *J. Hydraul. Eng.* 110:707.

Karcher, W., and Devillers, J. (1990) *Practical Applications of Quantitative Structure-Activity Relationships (QSAR) in Environmental Chemistry and Toxicology.* Karcher, W. and Devillers, J., Eds. Kluwer Academic Publisher, Dordrecht, Holland.

Kerler, F., and Schonherr, J. (1988) Accumulation of lipophilic chemicals in plant cuticles: prediction from octanol/water partition coefficients. *Arch. Environ. Contam. Toxicol.* 17:1–6.

Kier, L. B., and Hall, L. H. (1977) *Molecular Connectivity in Chemistry and Drug Research.* Academic Press, New York.

Kier, L. B., and Hall, L. H. (1986) *Molecular Connectivity in Structure-Activity Analysis:* Research Studies Press, Chichester, Hertfordshire, U.K.

Konemann, H. (1981) Quantitative structure-activity relationships in fish toxicity studies. Part I. Relationship for 50 industrial pollutants. *Toxicology* 19:229–238.

Larson, R. A., Hunt, L. L., and Blankenship, D. W. (1977) Formation of toxic products from a #2 fuel oil by photooxidation. *Environ. Sci. Technol.* 11:492–496.

Lemaire, J., Campbell, I., Hulpke, H., Guth, J. A., Merz, W., Philp, J., and von Waldow, C. (1982) An assessment of test methods for photodegradation of chemicals in the environment. *Chemosphere* 11:119–164.

Lewis, G. N. (1901) The law of physico-chemical change. *Proc. Amer. Acad. Sci.* 37, 49.

Ligocki, M. P., and Pankow, J. F. (1989) Measurements of the gas/particle distribution of atmopheric organic compounds. *Environ. Sci Technol.* 23:75–83.

Liss, P. S., and Slater, P. G. (1974) Flux gases across the air-sea interface. *Nature* 247:181–184.

Liu, D., Strachan, W. M. J., Thompson, K., and Ksawniewska, K. (1981) Determination of the biodegradability of organic compounds. *Environ. Sci. Technol.* 15:788–793.

Lyman, W. J., Reehl, W. F., and Rosenblatt, D. H. (1982) *Handbook of Chemical Property Estimation Methods,* McGraw-Hill, New York, NY.

Mabey, W., and Mill, T. (1978) Critical review of hydrolysis of organic compounds in water under environmental conditions. *J. Phys. Chem. Ref. Data* 7:383–415.

Mabey, W., Smith, J. H., Podoll, R. T., Johnson, H. L., Mill, T., Chou, T. W., Gates, J., Waight-Partridge, I., Jaber, H., and Vanderberg, D. (1982) Aquatic Fate Processes for Organic Priority Pollutants. EPA Report No. 440/4-81-014.

Mackay, D. (1979) Finding fugacity feasible. *Environ. Sci. Technol.*, 13:1218.

Mackay, D., and Leinonen (1975) The rate of evaporation of low solubility contaminants from water bodies. *Environ. Sci. Technol.* 9:1178–1180.

Mackay, D., and Shiu, W. Y. (1977) Aqueous solubility of polynuclear aromatic hydrocarbons. *J. Chem. Eng. Data* 22:399–402.

Mackay, D., Shiu, W. Y., and Sutherland, R. P. (1979) Determination of air-water Henry's law constants for hydrophobic pollutants. *Environ. Sci. Technol.* 13:333–337.

Mackay, D., Bobra, A., Shiu, W. Y., and Yalkowsky, S. H. (1980) Relationships between aqueous solubility and octanol-water partition coefficient. *Chemosphere* 9:701–711.

Mackay, D., and Shiu, W. Y. (1981) A critical review of Henry's law constants for chemicals of environmental interest. *J. Phys. Chem. Ref. Data* 10:1175–1199.

Mackay, D., and Paterson, S. (1981) Calculating fugacity. *Environ. Sci. Technol.* 15(9):1006–1014.

Mackay, D., and Paterson, S. (1982) Fugacity revisited. *Environ. Sci. Technol.* 16:654A–660A.

Mackay, D., Shiu, W. Y., Bobra, A., Billington, J., Chau, E., Yuen, A., Ng, C., and Szeto, F. (1982) Volatilization of organic pollutants from water, a report prepared for Environmental Research Laboratory, U.S. Environmental Protection Agency. Report EPA 60/S3-82-019 NTIS No. PB 82-230-939.

Mackay, D. (1982) Correlation of bioconcentration factors. *Environ. Sci. Technol.* 16:274–278.

Mackay, D. (1982) Effect of surface films on air-water exchange rates. *J. Great Lakes Res.* 8:299–306.

Mackay, D., Bobra, A. M., Chan, D., and Shiu, W. Y. (1982) Vapor pressure correlations for low volatility environmental chemicals. *Environ. Sci. Technol.* 16:645–649.

Mackay, D., and Yuen, A. T. K. (1983) Mass transfer coefficient correlations for volatilization of organic solutes from water. *Environ. Sci. Technol.* 17:211–216.

Mackay, D., Joy, M., and Paterson, S. (1983a) A quantitative water, air, sediment interaction (QWASI) fugacity model for describing the fate of chemicals in lakes. *Chemosphere* 12:981–997.

Mackay, D., Paterson, S., and Joy, M. (1983b) A quantitative water, air, sediment interaction (QWASI) fugacity model for describing the fate of chemicals in rivers. *Chemosphere* 12:1193–1208.

Mackay, D., and Paterson, S. (1983c) Indoor exposure to volatile chemicals. *Chemosphere* 12:143–154.

Mackay, D., Paterson, S., and Joy, M. (1983) Application of fugacity models to the estimation of chemical distribution and persistence in the environment, in *Fate of Chemicals in the Environment* Swann, R. L., and Eschenroeder, A., Ed., American Chemical Society Symposium Series 225:175–196.

Mackay, D., and Hughes, A. I. (1984) Three-parameter equation describing the uptake of organic compounds by fish. *Environ. Sci. Technol.* 18:439–444.

Mackay, D., Paterson, S., Cheung, B., and Neely, W. B. (1985) Evaluating the environmental behavior of chemicals with a level III fugacity model. *Chemosphere* 14:335–374.

Mackay, D., Paterson, S., and Schroeder, W. H. (1986) Model describing the rates of transfer processes of organic chemicals between atmosphere and water. *Environ. Sci. Technol.* 20:810–816.

Mackay, D., and Paterson, S. (1986) The fugacity approach to multimedia environmental modeling, in *Pollutants in a Multimedia Environment* Cohen, Y., Ed. Plenum Press, New York, NY, pp. 117–131.

Mackay, D., and Paterson, S. (1988) Partitioning models, in *Carcinogen Risk Assessment* Travis, C. C., Ed. Plenum Press, New York, NY, pp. 77–86.

Mackay, D., and Diamond, M. (1989) Application of the QWASI (Quantitative Water Air Sediment Interaction) fugacity model to the dynamics of organic and inorganic chemicals in lakes. *Chemosphere* 18:1343–1365.

Mackay, D. (1989) An approach to modelling the long term behavior of an organic contaminant in a large lake: Application to PCBs in Lake Ontario. *J. Great Lakes Res.* 15:283–297.

Mackay, D., and Stiver, W. H. (1989) The linear additivity principle in environmental modelling: Application to chemical behaviour in soil. *Chemosphere* 19:1187–1198.

Mackay, D., and Paterson, S. (1990) Fugacity models, in *Practical applications of quantitative structure-activity relationships (QSAR)* in *Environmental Chemistry and Toxicology* Karcher, W. and Devillers, J., Eds. Kluwer Acad. Publ., Dordrecht, Holland, pp. 433–460.

Mackay, D., and Paterson, S. (1991) Evaluating the multimedia fate of organic chemicals: A Level III fugacity model. *Environ. Sci. Technol.* 25:427–436.

Mackay, D., and Stiver, W. H. (1991) Predictability and environmental chemistry Chapter 8 pp. 281–297 in *Environmental Chemistry of Herbicides,* Vol. II. Grover, R. and Cessna, A. J., Eds. CRC Press, Boca Raton, FL.

McCall, P. J., Laskowski, D. A., Swann, R. L., and Dishburger, H. J. (1983) Estimation of environmental partitioning in model ecosystems. *Residue Res.* 85:231.

McLachlan, M., Mackay, D., and Jones, P. H. (1990) A conceptual model of organic chemical volatilization at waterfalls. *Environ. Sci. Technol.* 24:252–257.

Menzel, D. B. (1987) Physiological pharmacokinetic modeling. *Environ. Sci. Technol.* 21:944–950.

Merck Index. (1983) *An Encyclopedia of Chemicals, Drugs and Biologicals,* 10th ed. Windholz, M., Ed. Merck and Co., Inc., Rahway, NJ.

Mill, T., Mabey, W. R., Lan, B. Y., and Baraze, A. (1981) Photolysis of polycyclic aromatic hydrocarbons in water. *Chemosphere* 10:1281–1290.

Miille, M. J., and Crosby, D. G. (1983) Pentachlorophenol and 3,4-dichloroaniline as models for photochemical reactions in seawater. *Marine Chemistry* 14:111–120.

Miller, G. C., and Zepp, R. G. (1979) Effects of suspended sediments on photolysis rates of dissolved pollutants. *Water Res.* 13:453–459.

Miller, G. C., and Zepp, R. G. (1983) Extrapolating photolysis rates from the laboratory to the environment. *Residue Reviews* 85:89–110.

Miller, M. M., Wasik, S. P., Huang, G. L., Shiu, W. Y., and Mackay, D. (1985) Relationships between octanol-water partition coefficients and aqueous solubility. *Environ. Sci. Technol.* 19:522–529.

Mudambi, A. R., and Hassett, J. P. (1988) Photochemical activity of mirex associated with dissolved organic matter. *Chemosphere* 17:1133–1146.

National Academy of Sciences (1975) *Petroleum in the Marine Environment* National Academy Press, Washington, DC.

National Academy of Sciences (1985) *Oil in the Sea: Inputs, Fates and Effects* National Academy Press, Washington, DC.

Nazaroff, W. W., and Cass, G. R. (1986) Mathematical modeling of chemically reactive pollutants in indoor air. *Environ. Sci. Technol.* 20:924–934.

Nazaroff, W. W., and Cass, G. R. (1989) Mathematical modeling of indoor aerosol dynamics. *Environ. Sci. Technol.* 23:157–166.

Neely, W. B. (1979) Estimating rate constants for the uptake and clearance of chemicals by fish. *Environ. Sci. Technol.* 13:1506–1510.

Neely, W. B. (1980) *Chemicals in the Environment* M. Dekker, New York, NY.

Neely, W. B., and Mackay, D. (1982) Evaluative model for estimating environmental fate, in *Modeling the Fate of Chemicals in the Aquatic Environment*

Neely, W. B., and Blau, G. E. (1985) *Environmental Exposure from Chemicals*, Vols. I and II. CRC Press, Boca Raton, FL.

Neilson, A. H., Allard, A. S., and Remberger, M. (1985) Biodegradation and transformation of recalcitrant compounds, in *The Handbook of Environmental Chemistry*, Vol.2/Part C, Hutzinger, O., Ed. Springer-Verlag, Heidelberg, pp. 20–78.

Niimi, G. J., Veith, G. D., Regal, R. R., and Vaishnav, D. D. (1987) Structural features associated with degradable and persistent chemicals. *Environ. Toxicol. Chem.* 6:515–527.

Nirmalakhandan, N. N., and Speece, R. E. (1988a) Prediction of aqueous solubility of organic chemicals based on molecular structure. *Environ. Sci. Technol.* 22:323–338.

Nirmalakhandan, N. N., and Speece, R. E. (1988b) QSAR model for predicting Henry's constant. *Environ. Sci. Technol.* 22:1349⅓1357.

Nirmalakhandan, N. N., and Speece, R. E. (1989) Prediction of aqueous solubility of organic chemicals based on molecular structure. 2. Application to PNAs, PCBs, PCDDs, etc. *Environ. Sci. Technol.* 23:708–713.

Painter, H. A., and King, E. F. (1985) Biodegradation of water-soluble compounds, in *The Handbook of Environmental Chemistry*, Vol. 2/Part C, Hutzinger, O., Ed. Springer-Verlag, Heidelberg, pp. 87–118.

Pankow, J. F., and Morgan, J. J. (1981) Kinetics for the aquatic environment. *Environ. Sci. Technol.* 15:1306–1313.

Pankow, J. F. (1987) Review and comparative analysis of the theories on partitioning between the gas and aerosol particulate phases in the atmosphere. *Atmos. Environ.* 21: 2275–2283.

Pankow, J. F. (1988) The calculated effects of non-exchangeable material on the gas-particle distributions of organic compounds. *Atmos. Environ.* 22:1405–1409.

Paris, D. F., Steen, W. C., and Burns, L. A. (1982) Microbial transformation kinetics of organic compounds, in *The Handbook of Environmental Chemistry*, Vol. 2/Part B, Hutzinger, O., Ed. Springer-Verlag, Heidelberg, pp. 73–81.

Parke, D. V. (1982) The disposition amd metabolism of environmental chemicals by mammalia. *Handbook of Environmental Chemistry* Vol. 2/Part B, Hutzinger, O., Ed. Springer-Verlag, Heidelberg, pp. 141–178.

Paterson, S. (1985) Equilibrium models for the initial integration of physical and chemical properties, in *Environmental Exposure from Chemicals*, Vol. I, Neely, W. B., and Blau, G. E., Eds. CRC Press, Boca Raton, FL.

Paterson, S., and Mackay, D. (1985) The fugacity concept in environmental modelling, in *The Handbook of Environmental Chemistry*, Volume 2/Part C, Hutzinger, O., Ed. Springer-Verlag, Heidelberg, pp. 121–140.

Paterson, S., and Mackay, D. (1986) A pharmacokinetic model of styrene inhalation using the fugacity approach. *Toxicol. Appl. Pharmacol.* 82:444–453.

Paterson, S., and Mackay, D. (1987) A steady-state fugacity-based pharmacokinetic model with simultaneous multiple exposure routes. *Environ. Toxicol. Chem.* 6:395–408.

Paterson, S., and Mackay, D. (1989) A model illustrating the environmental fate, exposure and human uptake of persistent organic chemicals. *Ecolog. Modelling* 47:85–114.

Payne, J. R., and Phillips, C. R. (1985) Photochemistry of petroleum in water. *Environ. Sci. Technol.* 19:569–579.

Prausnitz, J. M., Lichtenthaler, R. N., and de Azevedo, E. G. (1986) *Molecular Thermodynamics of Fluid Phase Equilibria,* Second Edition. Prentice Hall, Englewood Cliffs, NJ.

Pruppacher, H. R., Semonin, R. G., and Slim, W. G. N. (Eds.) (1983) *Precipitation Scavenging, Dry Deposition and Resuspension. Vol. 1 Precipitation Scavenging and Vol. 2 Dry Deposition and Resuspension.* Elsevier, New York, NY.

Ramsey, J. C., and Andersen, M. E. (1984) A physiologically based description of the inhalation pharmacokinetics of styrene in rats and humans. *Toxicol. Appl. Pharmacol.* 73:159–175.

Reid, R. C., Prausnitz, J. M., and Poling, B. E. (1987) *The Properties of Gases and Liquids,* 4th ed. McGraw-Hill, New York, NY.

Reitz, R. H., and Gehring, P. J. (1982) Pharmacokinetic models, in *The Handbook of Environmental Chemistry,* Vol. 2/Part B, Hutzinger, O., Ed. Springer-Verlag, Heidelberg, pp. 179–195.

Rekker, R. F. (1977) *The Hydrophobic Fragmental Constant* Elsevier, Amsterdam/New York, NY.

Reuber, B., Mackay, D., Paterson, S., and Stokes, P. (1987) A discussion of chemical equilibria and transport at the sediment-water interface. *Environ. Toxicol. Chem.* 6:731–739.

Riederer, M. (1990) Estimating partitioning and transport of organic chemicals in the foliage/atmosphere system: discussion of a fugacity-based model. *Environ. Sci. Technol.* 24:829–837.

Roof, A. A. M. (1982) Basic principles of environmental photochemistry, in *The Handbook of Environmental Chemistry,* Vol. 2/Part B, Hutzinger, O., Ed. Springer-Verlag, Heidelberg, pp. 1–16.

Roof, A. A. M. (1982) Aquatic photochemistry, in *The Handbook of Environmental Chemistry,* Vol. 2/Part B, Hutzinger, O., Ed. Springer-Verlag, Heidelberg, pp. 43–70.

Rordorf, B. F. (1985) Thermodynamic properties of polychlorinated compounds: The vapour pressure and enthalpy of sublimation of ten dibenzo-p-dioxins. *Thermochemica Acta* 85:435–438.

Rordorf, B. F. (1987) Prediction of vapor pressures, boiling points and enthalpies of fusion for twenty-nine halogenated dibenzo-p-dioxins. *Thermochemica Acta* 112:117–122.

Sangster, J. (1989) Octanol-water partition coefficients of simple organic compounds, *J. Phys. Chem. Ref. Data* 18:1111–1229.

Satterfield, C. N. (1970) *Mass Transfer in Heterogeneous Catalysis* MIT Press, Cambridge, MA.

Sawhney, B. L., and Brown, K., Eds. (1989) *Reactions and Movements of Organic Chemicals in Soils* Soil Science Society of America, Special Publ. No. 22, Madison, WI.

Schramm, K. W., Reischl, A., and Hutzinger, O. (1987) A multimedia compartment model to estimate the fate of lipophilic compounds in plants. *Chemosphere* 16:2653–2663.

Seinfeld, J. H. (1975) *Air Pollution: Physical and Chemical Fundamentals.* McGraw-Hill Book Company, New York, NY.

Shaw, D. W., Ed. (1989) *IUPAC Solubility Data Series Vol. 37. Hydrocarbons with Water and Seawater. Part I. Hydrocarbons C_5 to C_7 Vol. 38. Hydrocarbons with Water and Seawater Part II. Hydrocarbons C_8 to C_{36}.* Pergamon Press, Oxford.

Sheehan, P., Korte, F., Kelin, W., and Boudreau, P. (1985) *Appraisal of Tests to Predict the Environmental Behaviour of Chemicals. SCOPE* 25 J. Wiley & Sons, New York, NY.

Sherwood, T. K., Pigford, R. L., and Wilke, C. R. (1975) *Mass Transfer* McGraw-Hill, New York, NY.

Shiu, W. Y., and Mackay, D. (1986) A critical review of aqueous solubilities, vapor pressures, Henry's law constants and octanol-water partition coefficients of the polychlorinated biphenyls. *J. Phys. Chem. Data* 15:911–929.

Shiu, W. Y., Doucette, W., Gobas, F. A. P. C., Andren, A., and Mackay, D. (1988) Physical chemical properties of chlorinated dibenzo-p-dioxins. *Environ. Sci. Technol.* 22:651–658.

Smith, J. H., Mabey, W. R., Bahonos, N., Holt, B. R., Lee, S. S., Chou, R. W., Bomberger, D. C., and Mill, T. (1978) Environmental Pathways of Selected Chemicals in Freshwater Systems: Parts I and II. Laboratory Studies. Interagency Energy-environment Research and Development Program Report. EPA-600/7-78-074. Environmental Research Laboratory Office of Research and Development. U.S. Environmental Protection Agency, Athens, GA, p. 304.

Sonnefeld, W. J., and Zoller, W. H. (1983) Dynamic coupled-column liquid chromatographic determination of ambient temperature vapor pressures of polynuclear aromatic hydrocarbons. *Anal. Chem.* 55:275–280.

Southwood. J. M., Harris, R. C., and Mackay, D. (1989) Modeling the fate of chemicals in an aquatic environment: The use of computer spreadsheet and graphics software. *Environ. Toxicol Chem.* 8:987–996.

Spacie, A., and Hamelink, J. L. (1982) Alternative models for describing the bioconcentration of organics in fish. *Environ. Toxicol. Chem.* 1:309–320.

Spencer, W. F., and Cliath, M. M. (1969) Vapor density of dieldrin. *Environ. Sci. Technol.* 3:670–674.

Spencer, W. F., Shoup, T. D., Cliath, M. M., Farmer, W. J., and Haque, R. (1979) Vapor pressures and relative volatility of ethyl and methyl parathion. *J. Agric. Food Chem.* 27:273–278.

Sposito, G. (1989) *The Chemistry of Soils,* Oxford University Press, New York, NY.

Stephen, H., and Stephen, T. (1963) *Solubilities of Inorganic and Organic Compounds,* Vols. 1 & 2, Pergamon, Oxford.

Streeter, H. W. and Phelps, E. B. (1925) A study of the pollution and natural purification of the Ohio River. U.S. Public Health Service Bulletin 146. Washington, DC.

Stumm, W., and Morgan, J. J. (1981) *Aquatic Chemistry,* John Wiley & Sons, New York, NY.

Suffett, I. H., and McCarthy, P. (1989) *Aquatic humic substances: Influence on fate and treatment of pollutants.* Advances in Chemistry Series 219. American Chemical Society, Washington, DC.

Suntio, L. R., Shiu, W. Y., and Mackay, D. (1988) A review of the nature and properties of chemicals present in pulp mill effluents. *Chemosphere* 17(7):1249–1290.

Suntio, L. R., Shiu, W. Y., Mackay, D., Seiber, J. N., and Glotfelty, D. (1988) Critical review of Henry's law constants for pesticides. *Rev. Environ. Contam. Toxicol.* 103:3–59.

Svenson, A., and Bjorndal, H., (1988) A convenient test method for photochemical transformation of pollutants in the aquatic environment. *Chemosphere* 17:2397–2405.

Thibodeaux, L. J. (1979) *Chemodynamics* Wiley-Interscience, New York, NY.

Thibodeaux, L. J., Reible, D. D., and Fang, C. S. (1986) Transport of Chemical Contaminants in the Marine Environment Originating from Offshore Drilling Bottom Deposits – A Vignette Model. *Pollutants in a Multimedia Environment,* Cohen, Y., Ed., Plenum Press, New York, NY, pp. 49–64.

Thomann, R. V., (1989) Bioaccumulation model of organic chemical distribution in aquatic food chains. *Environ. Sci. Technol.* 23:699–707.

Thompson, H. C., Jr., Kendall, D. C., Korfmacher, W. A., Rowland, K. L., Rushing, L. G., Chen, J. J., Kominsky, J. R., Smith, L. M. and Stalling, D. L. (1986) Assessment of the contamination of a multibuilding facility by polychlorinated biphenyls, polychlorinated dibenzo-p-dioxins, and polychlorinated dibenzofurans. *Environ. Sci. Technol.* 20:597–603.

Tinsley, I. J. (1979) *Chemical Concepts in Pollutant Behavior* Wiley-Interscience, New York, NY.

Travis, C. C. (1988) *Carcinogen Risk Assessment* Plenum Press, New York, NY.

Tuey, D. B., and Matthews, H. B. (1980) Distribution and excretion of 2,2′,4,4′,5,5′-hexabromobiphenyl in rats and man: Pharmacokinetic model predictions. *Toxicol. Appl. Pharmacol.* 53:420–431.

Valsaraj, K. T., Porter, J. L., Liljenfeldt, E. K., and Springer, C. (1986) Solvent sublation for the removal of hydrophobic chlorinated compounds from aqueous solutions. *Water Res.* 20:1161–1175.

Valsaraj, K. T. (1988) Binding constants for non-polar organics at the air-water interface: comparison of experimental and predicted values. *Chemosphere* 17:2049–2053.

Van Ness, H. C., and Abbott, M. M. (1982) *Classical Thermodynamics of Nonelectrolyte Solutions* McGraw-Hill Book Co., New York, NY.

Van Noort, P., Lammers, R., and Vondergem, E.. (1988) Rates of triplet humic acid sensitized photolysis of hydrophobic compounds. *Chemosphere* 17(1):35–38.

Veith, G. D., Defoe, D. L., and Bergstedt, B. V. (1979) Measuring and estimating the bioconcentration factor of chemicals in fish. *J. Fish. Res. Board Canad.* 36:1040–1048.

Veith, G. D., Call, D. J., and Brooke, L. T. (1983) Structure-toxicity relationships for the fathead minnow. *Pimephales promelas:* Narcotic industrial chemicals. *Canad. J. Fish. Aquat. Sci.* 40:743.

Verschueren, K. (1977) *Handbook of Environmental Data on Organic Chemicals.* Van Nostrand Reinhold, New York, NY.

Verschueren, K. (1983) *Handbook of Environmental Data on Organic Chemicals,* 2nd ed. Van Nostrand Reinhold, New York, NY.

Weast, R. C. (1984) *Handbook of Chemistry and Physics,* 64th ed., CRC Press, Boca Raton, FL.

Welling, P. G. (1986) *Pharmacokinetics: Processes and mathematics.* ACS Monograph 185. American Chemical Society, Washington, DC.

Westcott, J. W., Simon, C. G., and Bidleman, T. F. (1981) Determination of polychlorinated biphenyl vapor pressures by a semimicro gas saturation method. *Environ. Sci. Technol.* 15:1375–1377.

Whitman, W. G. (1923) The two-film theory of gas absorption. *Chem. Metal Eng.* 29:146–150.

Wolfe, N. L., Zepp, R. G., Paris, D. F., Baughman, G. L., and Hollis, R. C. (1977) Methoxychlor and DDT degradation in water: rates and products. *Environ. Sci. Technol.* 11:1077–1081.

Wolfe, N. L. (1980) Organophosphate and organophosphorothionate esters: application of linear free energy relationships to estimate hydrolysis rate constants for use in environmental fate assessment. *Chemosphere* 9:571–579.

Wolfe, N. L., Paris, D. F., Steen, W. C., and Baughman, G. L. (1980) Correlation of microbial degradation rates with chemical structures. *Environ. Sci. Technol.* 14:1143–1144.

Wong, A. S., and Crosby, D. G. (1981) Photodecomposition of pentachlorophenol in water. *Agric. Food Chem.* 29:125–130.

Wood, J. M. (1987) Chlorinated hydrocarbons: oxidation in the biosphere. *Environ. Sci. Technol.* 16:291A–296A.

Woodrow, J. E., Crosby, D. G., and Seiber, J. N. (1983) Vapor-phase photochemistry of pesticides. *Residue Reviews* 85:111–125.

Yalkowsky, S. H. (1979) Estimation of entropies of fusion of organic compounds. *Ind. Eng. Chem. Fundam.* 18:108.

Yalkowsky, S. H., Valvani, S. C., and Mackay, D. (1982) Estimation of the aqueous solubility of some aromatic compounds, *Residue Reviews,* 85:43–55.

Yalkowsky, S. H., Pinal, R., and Bannerjee, S. (1988) Water solubility: A critique of the solvatochromic approach. *J. Pharm. Sci.* 77:74–77.

Yalkowsky, S. H., Ed. (1989) *Arizona Data Base of Aqueous Solubility,* University of Arizona, Tucson, AZ.

Yamasaki, H., Kuwata, K., and Miyamoto, H. (1982) Effects of ambient temperature on aspects of airborne polycyclic aromatic hydrocarbons. *Environ. Sci. Technol.* 16:189–194.

You, F., and Bidleman, T. F. (1984) Influence of volatility on the collection of polycyclic aromatic hydrocarbon vapors with polyurethane foam. *Environ. Sci. Technol.* 18:330–333.

Zafiriou, O. C., Joussot-Dubien, J., Zepp, R. G., Zika, R. G. (1984) Photochemistry of natural waters. *Environ. Sci. Technol.* 18:358A–371A.

Zaidi, B. R., Stucki, G., and Alexander, M. (1988a) Low chemical concentration and pH as factors limiting the success of inoculation to enhance biodegradation. *Environ. Toxicol. Chem.* 7:143–151.

Zaidi, B. R., Murakami, Y., and Alexander, M. (1988b) Factors limiting success of inoculation to enhance biodegradation of low concentrations of organic chemicals. *Environ. Sci. Technol.* 22:1419–1425.

Zaidi, B. R., Murakami, Y., and Alexander, M. (1989) Predation and inhibitors in lake water affect the success of inoculation to enhance biodegradation of organic chemicals. *Environ. Sci. Technol.* 23:859–863.

Zepp, R. G., Wolfe, N. L., Gordon, J. A., and Baughman, G. (1975) Dynamics of 2,4-D esters in surface waters. Hydrolysis, photolysis, and vaporization. *Environ. Sci. Technol.* 9:1144–1150.

Zepp, R. G., and Cline, D. M. (1977) Rates of direct photolysis in aquatic environment. *Environ. Sci. Technol.* 4:359–366.

Zepp, R. G., and Baughman, G. L. (1978) Prediction of photochemical transformation of pollutants in the aquatic environment, in O. Hutzinger, I. H. von Lelyveld and B. C. S. Zoetman, Eds. *Aquatic Pollutants: Transformation and Biological Effects,* Pergamon Press, New York, NY, p. 237.

Zepp, R. G. (1980) Experimental approaches to environmental photochemistry, in O. Hutzinger, Ed. *Handbook of Environmental Chemistry* Springer-Vail, Berlin.

Zepp, R. G. (1982) Experimental approaches to environmental photochemistry, in *The Handbook of Environmental Chemistry,* Vol. 2/Part B, Hutzinger, O., Ed. Springer-Verlag, Heidelberg, pp. 19–41.

Zepp, R. G., Schlotzhauer, P. F., and Sink, R. M. (1985) Photosensitized transformations involving electronic energy transfer in natural waters: Role of humic substances. *Environ. Sci. Technol.,* 19:74–81.

Zwolinski, B. J., and Wilhoit, R. C. (1971) *Handbook of vapor pressures and heats of vaporization of hydrocarbons and related compounds.* API-44, TRC Publication No. 101, Texas A&M University, College Station, TX.

SYMBOLS

a	activity	H	Henry's Law Constant
A	area m²		
A	Clapeyron or Antoine Constant	I	Input rate mol/h
A	Constants in uptake efficiency equation	k	rate constant h⁻¹
		k	mass transfer coefficient m/h
B	diffusivity m²/h	K	partition coefficient
B	constants in Yamasaki or Clapeyron equations		
B	absorption coefficient	L	length m
B	area specific D values used in rivers mol/Pa m²	L	lipid content
		m	amount mol or g
		M	amount mol or g
C	concentration mol/m³ or g/m³	MP	melting point °C
C	constant in Antoine equation	N	flux mol/h
D	D value mol/h.Pa	P	pressure Pa
E	emission rate mol/h	Q	digestion coefficient
E	uptake efficiency	Q	scavenging ratio
f	fugacity Pa	R	gas constant 8.314 Pa.m³/mol K
F	fugacity ratio		
G	flow rate m³/h	Sc	Schmidt number

t	time h	X	concentration of sorbent kg/L
T	temperature °C or K	x	mole fraction in liquid
TSP	total suspended particulates $\mu g/m^3$ or ng/m^3	Z	Z value $mol/m^3.Pa$
U	velocity m/h	y	diffusion distance m
		Y	depth m
v	volume fraction		
v	molar volume m^3/mol		
V	volume m^3	γ	activity coefficient
		μ	chemical potential
W	width of river m	μ	viscosity Pa.S
		ϕ	fugacity coefficient
x	volume ratio	ρ	density kg/m^3

INDEX

absorption 190
acids 33, 96
activity coefficients 92, 95
advection 117, 121, 129, 145
aerosols 53, 82, 101, 188, 205
air 53, 92
air-water processes 75, 93, 163,
    165, 186, 188
alcohols 32
aldehydes 33
alkenes 31
aquivalence 107
aromatics 31
atmosphere 53, 122

bioaccumulation/bioconcentration
    36, 211
biodegradation 134
biomagnification 214
bioturbation 57
biphenyls 32
burial from sediments 58, 201,
    205

chemical potential 70
Clapeyron-Clausius 96
compartments 10

computer programs 111, 140,
    178, 182, 192, 198, 202, 216,
    233
concentration 7

deposition from air 54, 191, 205
deposition from water 57, 58,
    201, 205
diffusion 22, 61, 144, 145, 147,
    158, 161
diffusivity 147, 149, 153
dimensions 9
dispersed phases 83
D values 120, 128, 151, 166,
    175, 180, 205

eddy diffusion 153
EPA 26, 29, 30
equilibrium 19, 67
ethers 33
evaluative environments 26, 52, 64
evaporation 167, 190, 196, 205, 217
EXAMS 26, 175
exposure 219, 228

Ficks laws 147, 149
fish 56, 98, 211

food 220, 214
food chains 215
fugacity 2, 72, 87, 105, 119, 128
fugacity forms 112, 113, 139, 229

gas constant 8
gills 212
growth dilution 213

health 1
heat capacity 88
Henderson-Hasselbach 96
Henry's law 40, 75, 94, 165
heterogeneity 10
homogeneity 10
hydrocarbons 31, 41
hydrolysis 135
hydrophobicity 40

indoor conditions 216
ions 96, 106

ketones 33

lakes 203, 207
leaching 196
Level I 108
Level II 118, 119, 138
Level III 174
lipids 81, 199, 214
lipophilicity 40

mass balances 11, 138, 177, 206
mass transfer coefficients 151,
    162, 169, 171
metabolism 213
metals 34, 50, 106
Michaelis Menten 126
Michigan 30
Millington Quirk 159, 197
mineral matter 81, 195

nitrogen compounds 33
non-diffusive processes 24, 144

Occam 11
octanol-water 40, 45, 78, 98
OECD 30
Ontario 30
organic matter & carbon 55,
    59, 60, 62, 79, 187, 195,
    199
oxidation 137
oxygen 171, 210

PAHs 31
parallel resistances 172
particulate matter 55, 83
partition coefficients 69, 74, 85,
    118, 127
persistence 24, 35, 115
pesticides 34
pharmacokinetics 223
phenols 32, 33
photolysis 136
phthalate esters 34
plants 63
porous media 158, 160
prefixes 9, 27
pressure 8
priority chemicals 31
priority-selling 35, 39
pure solutes 102

QSARs 223
QWASI models 203, 208

rain 64, 191, 205
rate constants 17, 124, 126
reactions 123, 129, 133, 196,
    201, 205
residence time 24, 116, 130
resuspension 58, 201